P9-CSB-047

GENES AND INHERITANCE

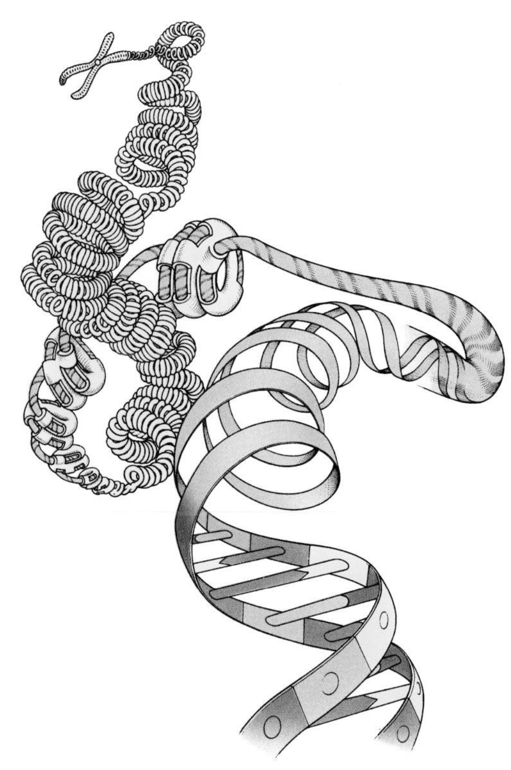

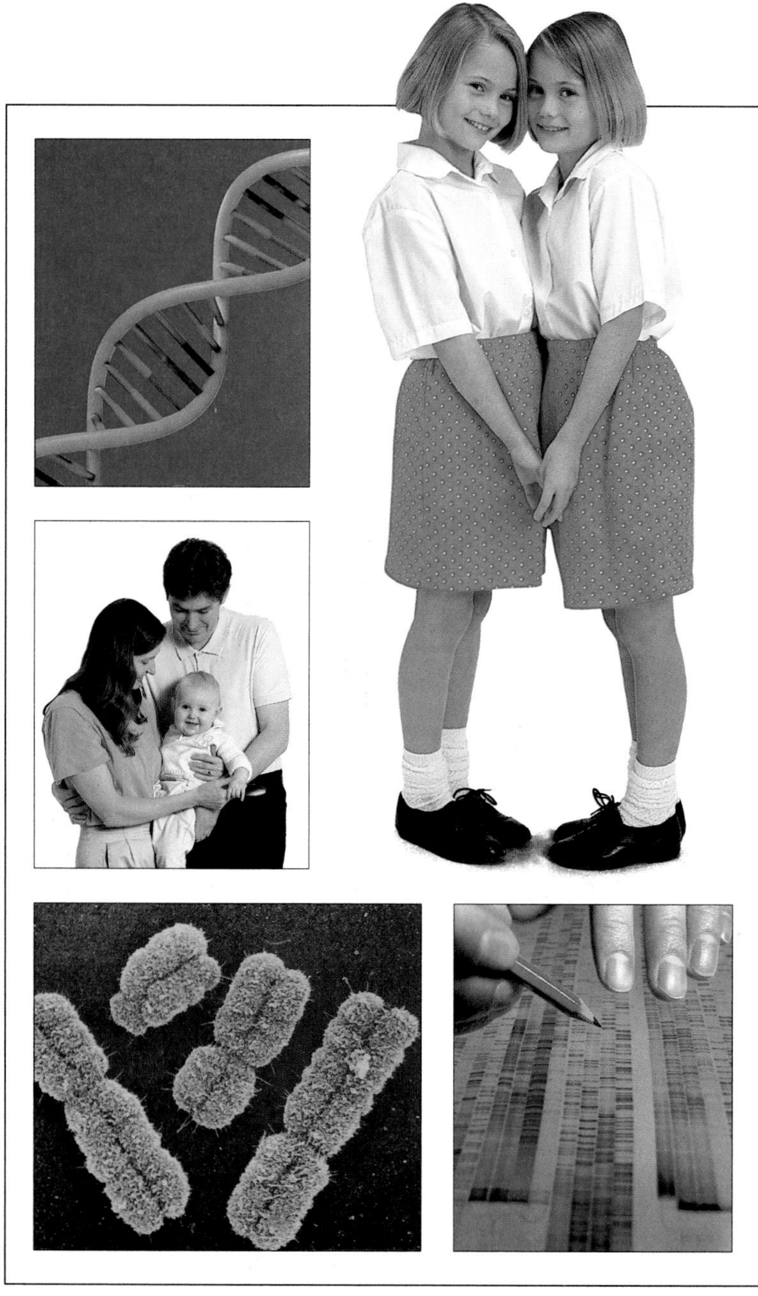

THE AMERICAN MEDICAL ASSOCIATION

GENES AND INHERITANCE

Medical Editor
CHARLES B. CLAYMAN, MD

THE READER'S DIGEST ASSOCIATION, INC.
Pleasantville, New York/Montreal

The AMA Home Medical Library was created and produced by Dorling Kindersley, Ltd., in association with the American Medical Association.

 Published in the United States by The Reader's Digest Association, Inc., Pleasantville, New York/Montreal

Manufactured in the United States of America

The information in this book reflects current medical knowledge. The recommendations and information are appropriate in most cases; however, they are not a substitute for medical diagnosis. For specific information concerning your personal medical condition, the AMA suggests that you consult a physician.

The names of organizations, products, or alternative therapies appearing in this book are given for informational purposes only. Their inclusion does not imply AMA endorsement, nor does the omission of any organization, product, or alternative therapy indicate AMA disapproval.

The AMA Home Medical Library is distinct from and unrelated to the series of health books published by Random House, Inc., in conjunction with the American Medical Association under the names "The AMA Home Reference Library" and "The AMA Home Health Library."

Library of Congress Cataloging in Publication Data

Genes and inheritance / medical editor, Charles B. Clayman.
p. cm. — (The American Medical Association home medical library)
At head of title: The American Medical Association.
Includes index.
ISBN 0-89577-460-7
1. Medical genetics—Popular works. 2. Human genetics—Popular works. I. Clayman, Charles B. II. American Medical Association. III. Series.
RB155.G3587 1993
616'.042—dc20 92-12671

FOREWORD

Nothing promises to change the field of medicine as much as the science of genetics. We are on the verge of developing new medical treatments and cures that only recently were not thought possible. In fact, until the middle of this century, little was known about our genes and how they work. Now we are learning that these units of hereditary information cause or contribute to our most common disorders, including cancer and heart disease, as well as to many rare conditions. In what promises to open up medicine's newest and potentially most fruitful frontier, scientists are learning how to supplement or replace disease-causing genes with healthy ones to cure even the most devastating inherited disorders.

This volume of the AMA Home Medical Library explains the genetic revolution in detailed yet understandable language. You will learn how DNA – the chemical substance that genes are made of – was discovered and the crucial role that genes play in our lives and health. In fact, your identity as an individual is a direct result of the interaction of your genes with the environment. Your genes largely determine your characteristics, your susceptibility or resistance to diseases, your body's reaction to medications, whether you carry and can transmit a genetic disease to your children, or whether you have or will develop an inherited disorder yourself.

In this volume, we explain how genes can become altered to cause diseases of all kinds, and we describe the sophisticated new laboratory tests that can detect many hereditary disorders even before birth. We hope this volume will help you and your family understand how doctors can predict your risk of developing a genetic disease and how doctors can, in many cases, help you diminish those risks.

James S. Todd MD

JAMES S. TODD, MD
Executive Vice President
American Medical Association

THE AMERICAN MEDICAL ASSOCIATION

CONTENTS

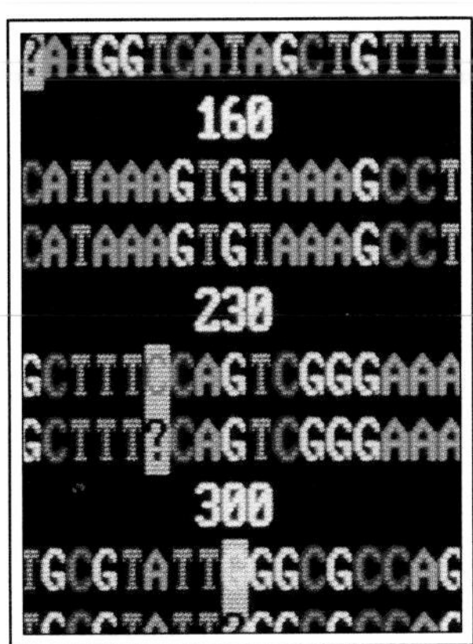

160
CATAAAGTGTAAAGCCT
CATAAAGTGTAAAGCCT
230
GCTTT?CAGTCGGGAAA
300

CHAPTER ONE

WHAT ARE GENES?

It is often easy to see that children resemble their parents – not only in appearance, but in personality traits and even susceptibility to certain diseases. But it has only been in the last 150 years that science has uncovered the molecular basis for these observations. Applying this knowledge to the understanding and treatment of illness began only recently.

The first part of this chapter traces the path that scientists took toward understanding heredity. Long before they had the technology to examine the detailed makeup of cells, scientists suspected that what people inherit from their parents is controlled by specific combinations of chemicals that direct biological processes. When they watched cells divide under a microscope and saw threadlike structures called chromosomes, they reasoned that the chromosomes were somehow involved in transmitting characteristics from one generation to the next. Because they observed an abundance of human characteristics and relatively few chromosomes, researchers deduced that each chromosome contained distinct hereditary units responsible for those characteristics. They called those units "genes."

A major scientific breakthrough in understanding heredity was the discovery of the structure of the principal chemical compound that makes up chromosomes – deoxyribonucleic acid (DNA). DNA is present in all living cells and carries all the information required for life. You might call it an instruction manual for life processes. In humans, this instruction manual exists in the form of 46 "chapters," or chromosomes, half of which are contributed by each parent.

Genes can be thought of as sentences that each carry an instruction for a specific characteristic (such as eye color) or function (such as hormone production).

Although it is our genes that make us unique, the similarities among us are greater than the differences. Because all forms of life have evolved over billions of generations from common ancestors, many of our genes are identical to those of other animals and even to plants. Chimpanzees share most of our genes, cows share many, carrots and cabbages share a few. The difference lies in how the genes are put to work.

The influence of genes on some traits, such as intelligence and behavior, is difficult to determine. Your identity as an individual results from the complex interaction of your genes and your environment.

A SHORT HISTORY OF GENETICS

Farmers and animal breeders have long relied on heredity to improve the quality of plants and animals by crossbreeding. Until recently, however, people had little idea how the complex mechanisms of heredity worked. The science of genetics is now providing answers. Genetics is being used to solve medical mysteries and to diagnose and treat illness. In the near future doctors hope to use the new knowledge to cure and even to prevent many common genetic disorders.

Homunculus
At the time of the early Greeks, many people believed that each sperm contained a miniature person called a homunculus that entered a woman's uterus and grew there until it was born. It was not until the middle of the 19th century that studies of embryos disproved this idea and scientists performed experiments that revolutionized our understanding of heredity.

THEORIES OF EVOLUTION

For centuries, people believed that all species of plants and animals were created at about the same time. However, fossil studies have shown that most species on Earth evolved over hundreds of millions of years; that is, their structure changed gradually over generations. These gradual alterations are responsible for the astonishing diversity and countless number of species alive today, including our own. Since the early 19th century, scientists have debated how these changes occurred.

Acquired characteristics

In the early 19th century, French biologist Jean Baptiste Lamarck (1744-1829) introduced the idea that plants and animals changed or adapted as they reacted to changes or new threats in their environment. Lamarck theorized that the characteristics that an organism altered or acquired during its life were transmitted to offspring. Widely accepted at the time, Lamarck's ideas were later discounted in favor of Darwin's theory of evolution by natural selection.

Fossil studies
Fossil studies continue to provide us with information about the plants and animals that have lived on Earth over hundreds of millions of years. The fish at left, called a coelacanth, was known only from fossils and thought to be extinct until living descendants (below) with slightly altered characteristics were found earlier this century.

THE GALÁPAGOS FINCHES

While visiting the Galápagos Islands off South America, Darwin observed that different species of finches on different islands resembled a single species of finch on the South American mainland. Darwin thought it likely that the finches had colonized the islands from the mainland many years before and had then been isolated. He later postulated that a variety of species evolved over many generations on the different islands as the birds adapted to distinctly different environments.

Natural selection

In 1859, British naturalist Charles Darwin (1809-1882), in his book *The Origin of Species,* presented his theory of the way in which species originated and evolved through generations. From his careful observation of plants and animals, Darwin realized that the members of any species vary a little in their characteristics and that, because of this variation, some members are better adapted than others to their environment. This led him to suggest that gradual changes occur in a species over generations because better adapted members survive longer and produce more offspring than those who are less well adapted. In Darwin's theory of "survival of the fittest," those organisms that are "fit" (better adapted) would be "naturally selected" over organisms with less helpful traits. They would then pass those advantageous traits on to offspring, who would also be more likely to survive and reproduce and pass on the same beneficial traits. In this way, the traits that help a species adapt to its environment would gradually become more common. Conversely, the traits that reduce an organism's chances for survival and reproduction would gradually become rare within a species over generations.

THE LAWS OF HEREDITY

Long before scientists understood the molecular basis of heredity, a part-time scientist established some of the basic rules. From his studies on the garden pea, Gregor Mendel (1822-1884), an Austrian monk and teacher, formulated an accurate account of how genetic characteristics are transmitted from one generation to the next (see page 12). His work led to the new field of genetics and even today provides fresh insights into the important role that genes play in our lives.

Darwin's unifying theory
In his theory of evolution, Charles Darwin (left) offered the first plausible scientific explanation of how life – in its awesome variety and complexity – evolved on this planet. He provided a unifying concept into which all subsequent discoveries in genetics and other fields of biology could be integrated.

WHY DO GIRAFFES HAVE LONG NECKS?

Lamarck's theory suggested that giraffes developed long necks because they had to stretch to feed on the leaves of trees; the acquired characteristic of a long neck was then passed on to offspring. According to Darwin's theory, those ancestors of today's giraffes that were born with longer necks were better equipped to eat the leaves off the treetops and, therefore, better able to survive and reproduce. Their offspring were more likely to have long necks and, in turn, survive and reproduce.

Beak variation
Darwin found that the size and shape of the beaks of different finches were related to their diets. The medium ground finch (top row left), which feeds mainly on large seeds, has a short, heavy beak. The cactus ground finch (bottom row right) has a long beak designed for probing flowers of the prickly pear, on which it feeds.

MENDEL: THE FIRST GENETICIST

Gregor Mendel (left), like his contemporary Charles Darwin, realized that species of plants and animals differed considerably in characteristics such as size, shape, and color. Mendel decided to find out what determined a species' traits and how those traits were inherited. He chose the garden pea as his research subject. Mendel studied seven of the plant's traits, each of which had two distinct forms – the stems, for example, were either tall or short.

Seven pairs of characteristics
Mendel studied pea plants to determine how certain characteristics are inherited. Of the seven traits that Mendel studied, those he called dominant are shown in the left column below; the corresponding characteristics that he called recessive are in the right column.

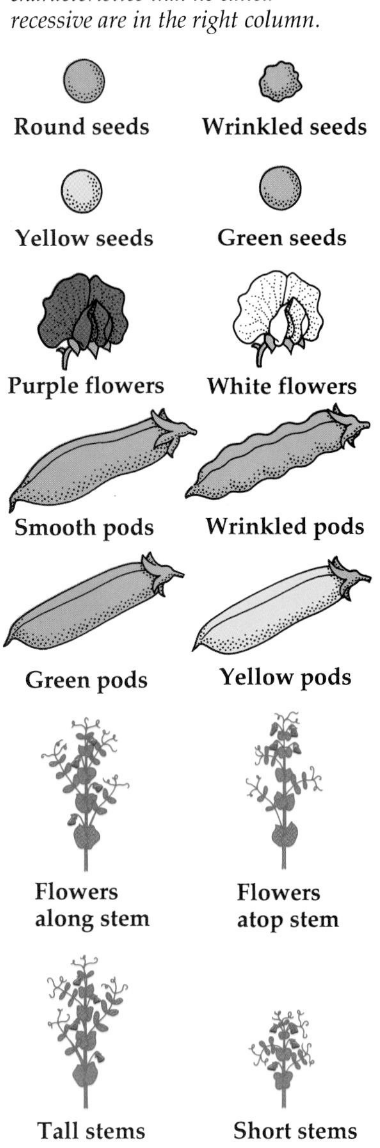

1 In a process called purebreeding, Mendel established strains of pea plants for each trait that he wanted to study. For example, he established a strain of tall plants, and a strain of short plants. Mendel then crossbred plants from these pure strains. He crossed a tall plant with a short plant (right), and a plant that produced round seeds with a plant that produced wrinkled seeds (below right).

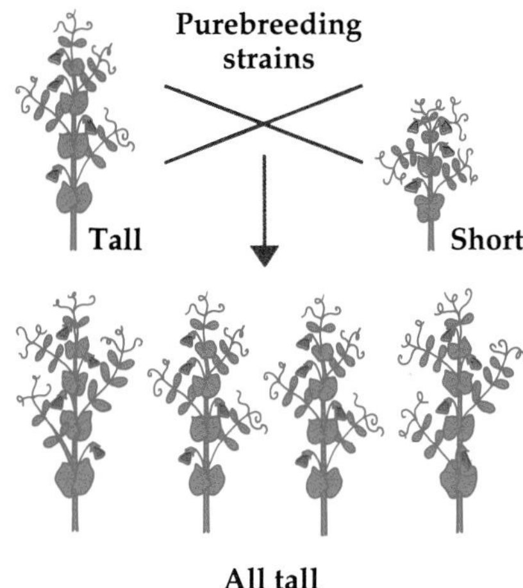

2 Mendel found that offspring (called hybrids) from these crossbreedings never exhibited a blend of characteristics. Instead, they displayed just one of the traits. For example, all the offspring of the crosses between tall and short plants were tall; all the offspring of the crosses between plants with round and wrinkled seeds had round seeds. Mendel called the persistent traits, such as tall stems and round seeds, "dominant." The forms that seemed to have disappeared, such as short stems and wrinkled seeds, he called "recessive."

Purebreeding strains

Round seeds

Wrinkled seeds

All round

3 Mendel then took the hybrid plants produced from his first generation of crossbreedings and crossed them with each other. The offspring of this second generation of crosses (right) followed two remarkable patterns. First, some of the offspring exhibited the recessive trait (such as a short stem), which had not appeared in the first generation. Second, for every one offspring with the recessive trait (a short stem), there were three plants with the dominant trait (a tall stem).

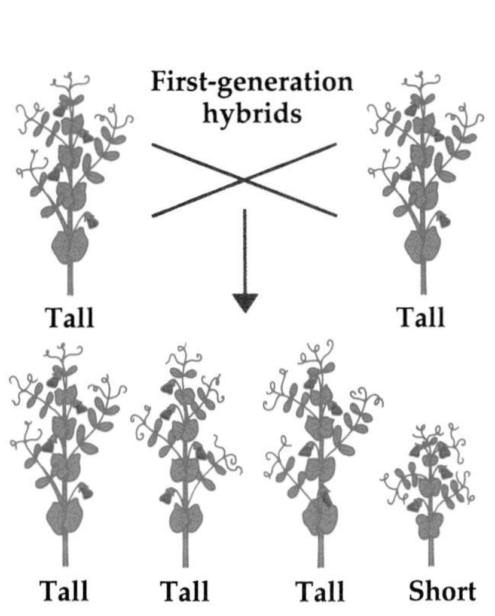

4 Mendel's experiments are illustrated below with the heredity of tall and short stems. He theorized that each of the traits he studied was determined by the interaction of a pair of factors (later called genes) that can come in one of two forms. Mendel referred to the alternate forms of the factors that determine stem length as T (for long stems) and t (for short stems). He determined that, if both forms are present in a plant, one form (T) is dominant and it masks the effect of the other form (t), which is recessive.

5 The purebred tall plants had the factors TT; purebred short plants had the factors tt. The offspring of the first round of crosses were all Tt but, because T was dominant, the plants all had long stems. Each plant passes on only one of its two factors to each offspring. When the first-generation hybrids were crossed with each other, an average of one in four offspring inherited the recessive (t) factor from both parent plants and, therefore, had short stems. The others inherited at least one dominant (T) factor, giving them long stems.

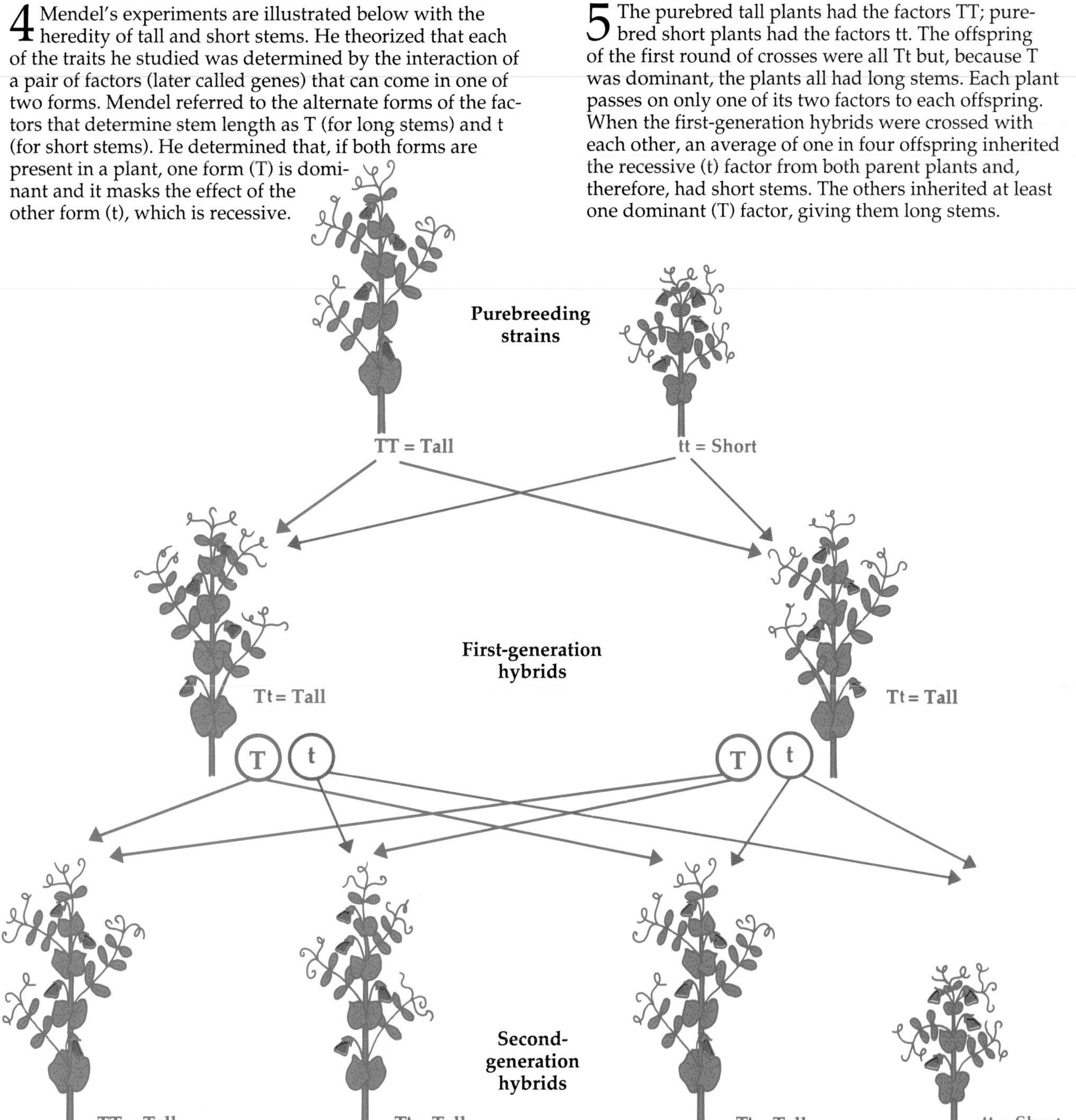

WHY WERE MENDEL'S PEAS IMPORTANT?

Through his experiments, Mendel established the following fundamental principles of heredity:

◆ The inherited traits of a living organism are determined by distinct factors (now called genes) that are transmitted from one generation to the next.

◆ Many characteristics of organisms are determined by the interaction of a pair of hereditary factors.

◆ The factors in a pair determining a trait may be the same or different. When they are different, the dominant form masks the effect of the other form, which is called recessive.

◆ Organisms with specific traits are produced in predictable numbers according to statistical rules from one generation to the next.

Mendel's rules form the foundation of the science of genetics and our understanding of how traits are inherited.

THE DISCOVERY OF CHROMOSOMES

In 1882, German anatomist Walther Flemming (1843-1905) was experimenting with synthetic dyes that would enable him to examine transparent animal cells under a microscope. He noticed that some parts of the cell absorbed the dyes while other parts did not. The dyes were absorbed most by a defined area inside the cell called the nucleus. Flemming named the stained material "chromatin," from the Greek word for color. He observed that the structure of the chromatin changed when cells divided to form two identical daughter cells. At the beginning of the process of cell division, the threadlike material condensed into a number of short, compact segments. Flemming's colleague Heinrich Wilhelm Waldeyer (1836-1921) later named these compact segments chromosomes, a term meaning "colored bodies."

Cells and chromosomes
In 1882, after Walther Flemming first observed chromosomes inside a cell (below, magnified 1,500 times), Edouard van Beneden, a Belgian cell expert, showed that each cell in animals of the same species contained the same number of chromosomes. He also found that sperm and egg cells contained exactly half the number of chromosomes present in other cells.

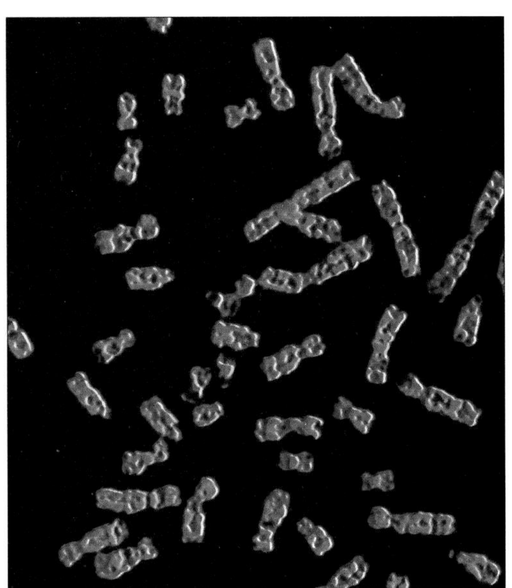

Purple eyes | Red eyes | Brown eyes | White eyes

Chromosomes and genes
In the 1930s, zoologist Thomas Morgan used fruit flies in his genetics experiments because they breed rapidly. By studying the heredity of a variety of characteristics such as eye color, Morgan proved that chromosomes are the carriers of inherited factors and that these factors – genes – are arranged in linear form along the threadlike chromosomes.

DISCOVERING THE GENE

After the discovery of chromosomes, many scientists suspected that chromosomes were the carriers of the hereditary factors that Gregor Mendel had studied several decades before in his experiments with peas. However, it was unclear how the 40 to 50 chromosomes then estimated to be contained in human cells could account for the far greater number of human characteristics. We now know that people have 46 chromosomes – half from each parent – arranged in 23 pairs. By the early 1900s scientists suggested that each chromosome carries many distinct hereditary units that determine the characteristics of a plant or animal. They called those units genes. At the time, no one had any idea of what a gene actually looked like. Nevertheless, the notion that a gene is the hereditary factor responsible for a single characteristic or function became widely accepted.

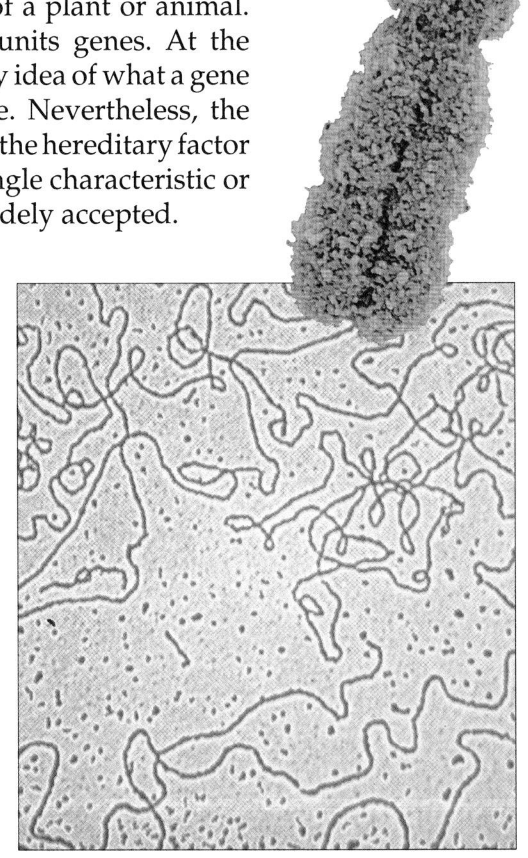

Chromosomes and DNA
Early geneticists understood that chromosomes contained genetic material, but knew little about how this material was arranged. We now know that chromosomes are threadlike strands of DNA (right, magnified 50,000 times). Just before cells begin to divide to make copies of themselves, the individual strands of DNA become tightly coiled into the compact X shape of chromosomes (above right, magnified 10,000 times).

SICKLE CELL TRAIT

Sickle cell anemia, which occurs mainly in blacks, is an inherited blood disease caused by an abnormality in the structure of hemoglobin, the oxygen-carrying pigment in red blood cells. The disorder occurs in people who inherit the gene mutation from both of their parents. Sickle cell anemia is no longer always fatal in childhood, but it still causes serious medical problems. If the disease gene is inherited from only one parent, most of the hemoglobin in the red blood cells is normal and the person has a condition called sickle cell trait, which causes few health problems. The sickle cell trait appears to provide some protection against falciparum malaria; scientists believe this is why the mutant sickle cell gene is common in parts of Africa where malaria is present.

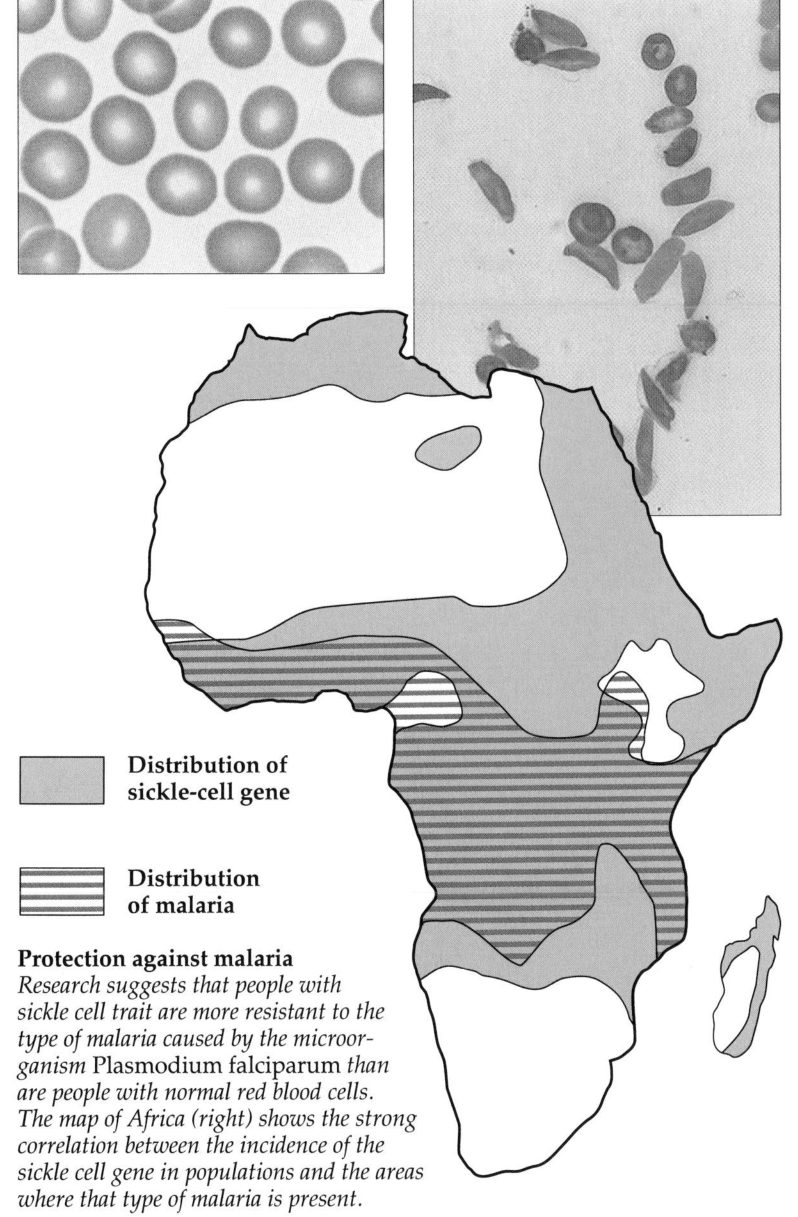

Protection against malaria
Research suggests that people with sickle cell trait are more resistant to the type of malaria caused by the microorganism Plasmodium falciparum *than are people with normal red blood cells. The map of Africa (right) shows the strong correlation between the incidence of the sickle cell gene in populations and the areas where that type of malaria is present.*

ASK YOUR DOCTOR MUTATIONS

Q My son has been diagnosed as having hemophilia. I thought hemophilia was hereditary, but nobody in our family has ever had it. How could this happen?

A Hemophilia is caused by a gene carried on the X sex chromosome. It almost always occurs in males, who inherit the harmful gene on their mother's X chromosome. Most affected families have a history of the disease. But about one third of the time, hemophilia results from a new mutation that occurred during the formation of a parent's egg or sperm cell. This new mutation may be transmitted, in the next generation or a subsequent generation, to a boy who will have the disorder.

Q Can new gene mutations be detected during pregnancy?

A Usually not. Because new gene mutations cannot be predicted, it is not possible to identify those pregnancies that may be at risk. In addition, a large number of different genes can mutate and cause genetic disorders; this makes it impossible to test during pregnancy for every single disorder that might result from a new gene mutation.

Q How often do gene mutations occur in our cells?

A Mutations that have any noticeable or harmful effect are relatively rare. The mutation rates of different genes vary, but the estimated risk that an individual gene will mutate during the formation of an egg or sperm cell ranges from one in 100,000 to one in a million. Considering the complexity of the DNA copying process, this is a remarkably low error rate.

PATTERNS OF HEREDITY

HEREDITY IS THE TRANSMISSION of characteristics from parents to children through the influence of genes. The basic laws of heredity were first described in the 1860s, long before the chemical nature of genes was known. Our understanding of the complexity of genes and their role in heredity advanced rapidly after the discovery in 1953 of the structure of DNA (deoxyribonucleic acid), the fundamental carrier of genetic information.

The genes in a cell are packaged in long strands of DNA called chromosomes. Inside a chromosome, genes are arranged in a precise order. The genes controlling most characteristics come in pairs – one inherited from the father and one from the mother. The paired genes are located at corresponding sites on 22 pairs of chromosomes called autosomes. The sex chromosomes X and Y make up the remaining chromosome pair. Females have two X sex chromosomes; males have one X and one Y. Every one of your cells contains a complete set of all of your genes. Each set is packaged in 23 pairs of chromosomes – 46 chromosomes in all.

The heredity of some traits, such as eye color, is determined by a single pair of genes. Other, more complex traits, such as intelligence, result from combinations of various genetic and environmental influences.

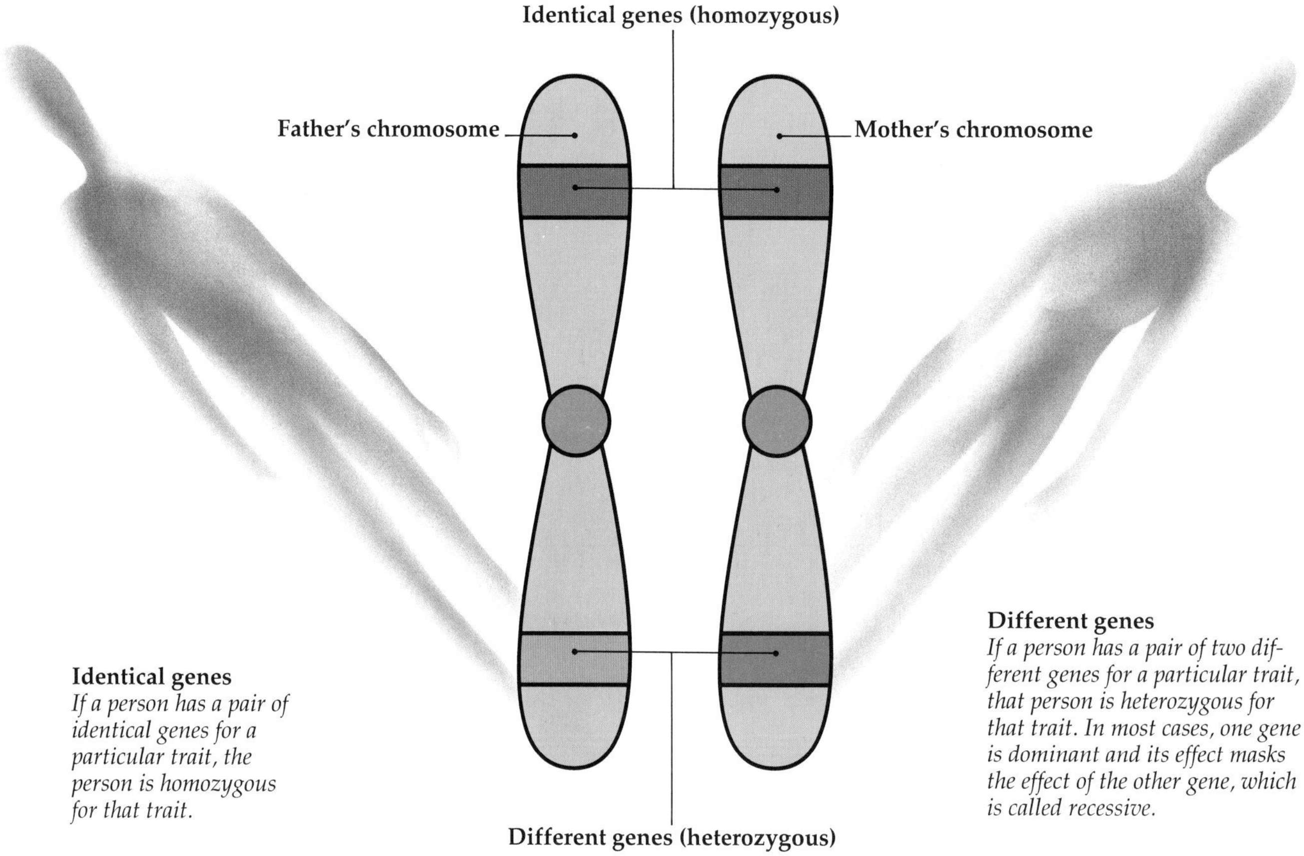

Identical genes (homozygous)

Father's chromosome

Mother's chromosome

Different genes (heterozygous)

Identical genes
If a person has a pair of identical genes for a particular trait, the person is homozygous for that trait.

Different genes
If a person has a pair of two different genes for a particular trait, that person is heterozygous for that trait. In most cases, one gene is dominant and its effect masks the effect of the other gene, which is called recessive.

DOMINANT AND RECESSIVE CHARACTERISTICS

Many traits are determined by a single pair of genes. The members of the pair may be either dominant or recessive. Because a dominant gene overrides a recessive gene, only one dominant gene is required to produce its effect on a person. But a person must have two copies of a recessive gene to exhibit the characteristic determined by that gene. For example, in simplified terms, eye color is regulated by a single pair of genes, either of which may be a gene for blue eyes or a gene for brown eyes. The gene for brown eyes is dominant and the gene for blue eyes is recessive. Therefore, a person with brown eyes may have either one or two copies of the dominant gene for brown eyes. A person has blue eyes only if he or she has inherited two copies of the recessive gene for blue eyes.

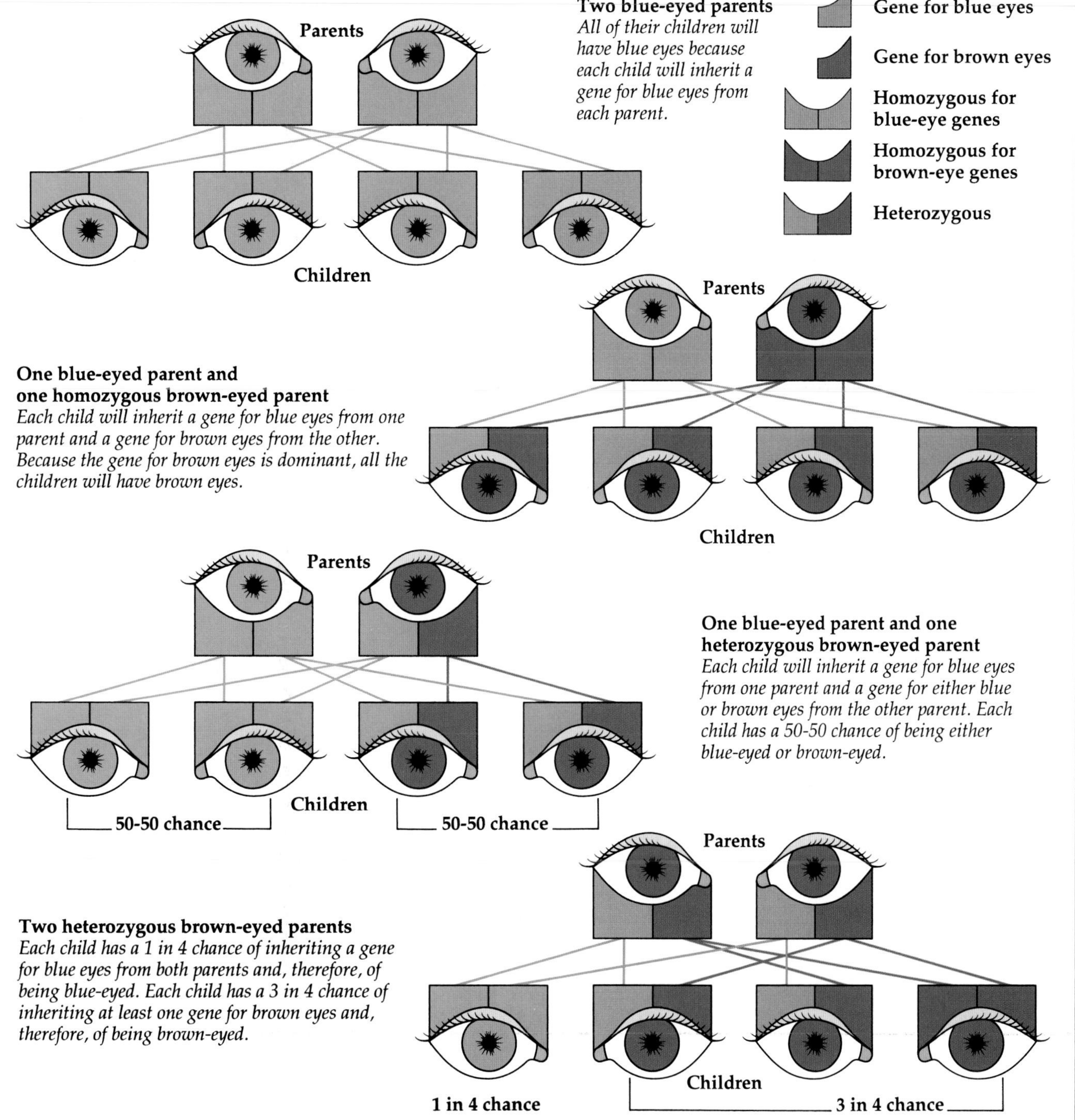

Two blue-eyed parents
All of their children will have blue eyes because each child will inherit a gene for blue eyes from each parent.

One blue-eyed parent and one homozygous brown-eyed parent
Each child will inherit a gene for blue eyes from one parent and a gene for brown eyes from the other. Because the gene for brown eyes is dominant, all the children will have brown eyes.

One blue-eyed parent and one heterozygous brown-eyed parent
Each child will inherit a gene for blue eyes from one parent and a gene for either blue or brown eyes from the other parent. Each child has a 50-50 chance of being either blue-eyed or brown-eyed.

Two heterozygous brown-eyed parents
Each child has a 1 in 4 chance of inheriting a gene for blue eyes from both parents and, therefore, of being blue-eyed. Each child has a 3 in 4 chance of inheriting at least one gene for brown eyes and, therefore, of being brown-eyed.

SIMPLE INHERITANCE

Some characteristics, such as blood type, the color of eyes and hair, and even the ability to taste certain substances, are controlled by a single pair of genes located on autosomes (nonsex chromosomes). The transmission of these traits follows the basic laws of heredity first described by Gregor Mendel in his experiments with pea plants in the 1860s (see page 12). The inheritance of hair color, for example, is usually determined by a single pair of genes. The hair-color genes can be in either of two principal forms, one for dark hair and the other for light hair. When a person has a copy of each of the two forms, the instructions provided by the gene for dark hair (the dominant gene) override the instructions provided by the gene for light hair (the recessive gene). To be dark-haired, a person needs to inherit only one gene for dark hair from one parent, while, to be light-haired, he or she must inherit two copies of the gene for light hair, one from each parent. As a result, dark hair is considered the dominant trait and light hair is the recessive trait. Similar rules apply to many other characteristics that are controlled by a single pair of genes, including those shown on this page and on page 51.

Curly hair
Like dark hair, curly hair is a dominant trait. A person with curly hair may have only one dominant gene for curly hair and a recessive gene for straight hair, or two copies of the dominant curly hair gene.

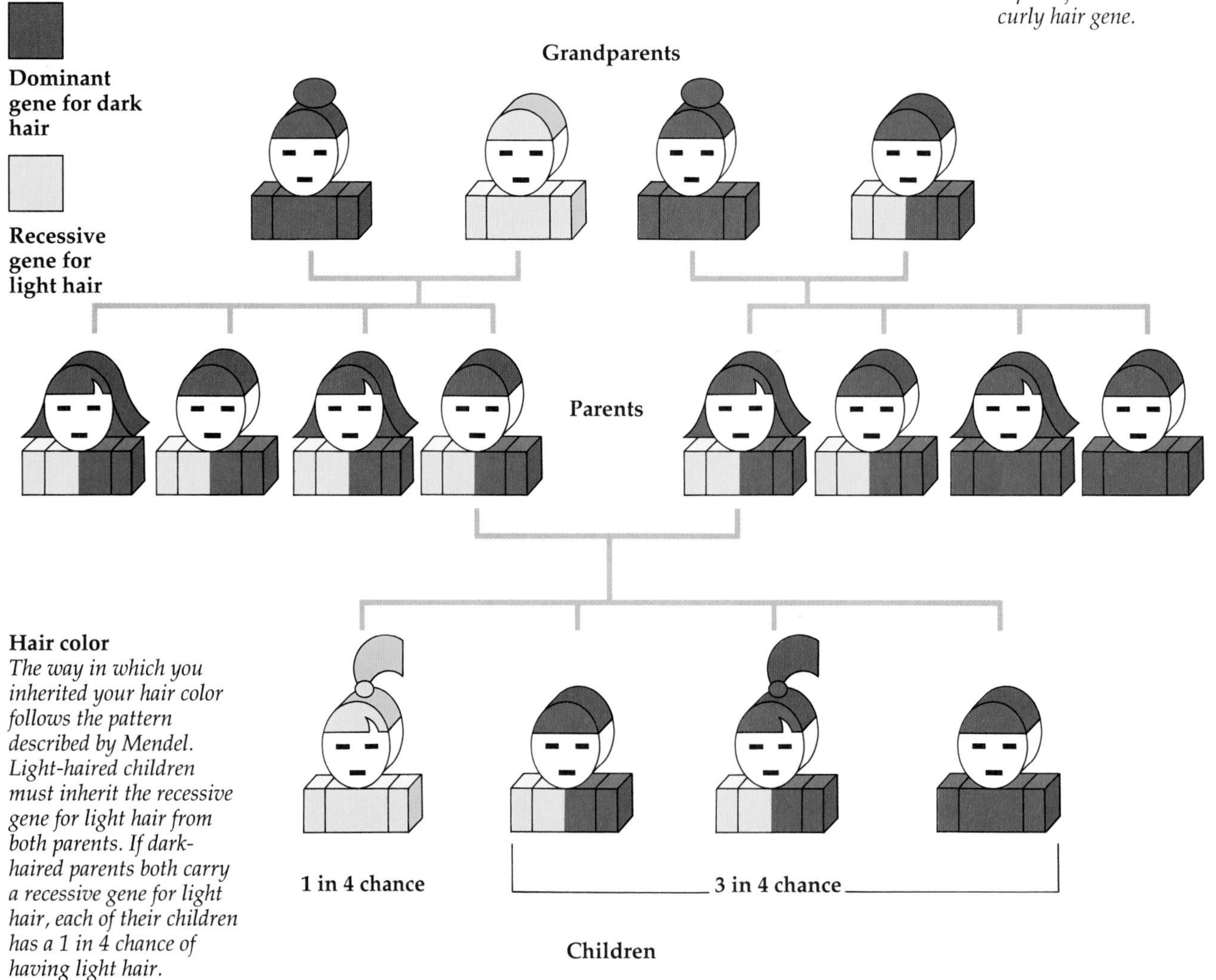

Hair color
The way in which you inherited your hair color follows the pattern described by Mendel. Light-haired children must inherit the recessive gene for light hair from both parents. If dark-haired parents both carry a recessive gene for light hair, each of their children has a 1 in 4 chance of having light hair.

Albinism
Albinism, a condition that results from the body's inability to make sufficient amounts of a pigment called melanin, is a good example of a recessive disorder. Albinos have little or no pigment in their skin and hair, making them extremely sensitive to the damaging effects of the sun. An affected person has inherited two copies of the recessive gene for albinism, one from each parent.

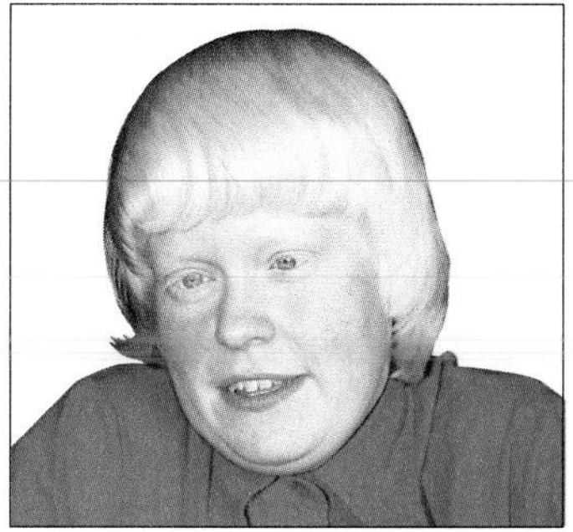

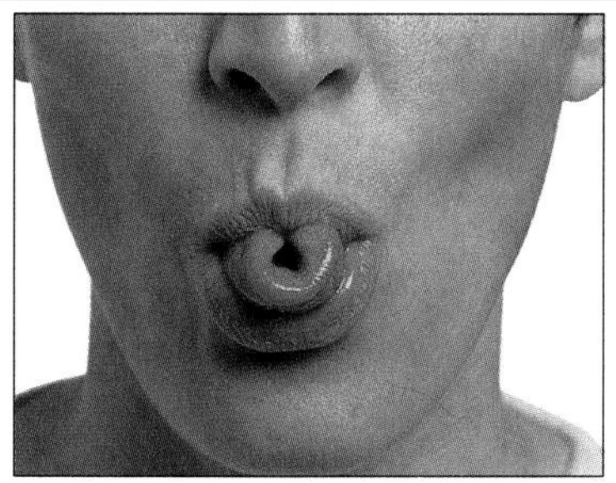

Ability to roll your tongue
The ability to roll your tongue (left) is determined by a dominant gene that controls certain muscles in your tongue. If you have either a single or double copy of the gene, as do about 85 percent of people, you can roll your tongue easily. If you do not have at least one copy of the gene, you will be unable to roll your tongue.

Taste
Geneticists think that the ability to taste some substances may be inherited. In an accidental discovery, scientists found that a chemical compound called phenylthiocarbamide can be tasted only by people who have one or two copies of a particular dominant gene that makes the chemical taste bitter. More than 70 percent of Americans can taste the chemical.

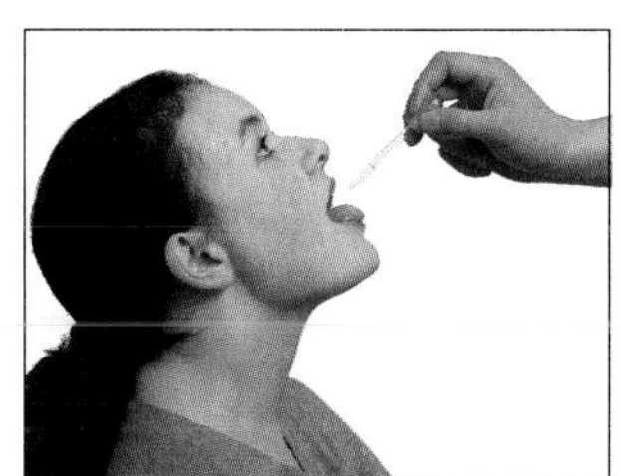

FAMILY INHERITANCE PATTERNS

The pattern of inheritance of a trait in a family from generation to generation can help doctors to identify hereditary disorders and the way in which they are transmitted from parents to child. The example below shows the inheritance pattern in one family of a dominant disorder called polydactyly. A person with polydactyly is born with an extra finger or toe.

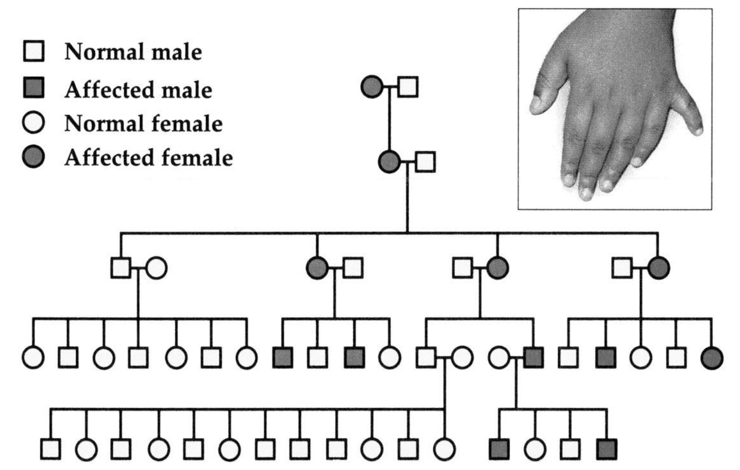

ASK YOUR DOCTOR HEREDITY

Q Both my wife and I have dark hair; our daughter has light hair. Since there is a 1 in 4 chance of our having a light-haired child, does this mean that any other children we have will be dark-haired?

A No. The statistic applies to every child that you have. Each of your children has a 1 in 4 chance of being light-haired, regardless of the hair color of any previous children. Your next child may have either light hair or dark hair, but there is a significantly greater chance (75 percent) that he or she will have dark hair.

Q My husband and I both have brown eyes and, so far, we have two brown-eyed children. Does this mean we will never have any blue-eyed children?

A It is impossible to say for sure without knowing the eye color of your parents and your wife's parents. In addition to the dominant brown-eye gene, you and your wife may each carry the recessive gene for blue eyes. If that is the case, each of your children has a 1 in 4 chance of having blue eyes.

Q Our child is an albino but my wife and I are not. If this condition is inherited, how could this have happened?

A Because albinism is a recessive disorder, a child must inherit a copy of the recessive gene from both parents to be noticeably affected by it. This means that both you and your wife carry the gene for albinism. In your cases, the recessive gene has been neutralized by the presence of a normal, dominant gene. But each child that you have has a 1 in 4 chance of being an albino.

HOW IS BLOOD TYPE INHERITED?

Using the ABO blood-grouping system, blood type is classified by the presence or absence of different types of molecules, called antigens, on the surface of red blood cells. People whose cells carry type A antigens have type A blood; people with type B antigens have type B blood. If their cells carry both A and B antigens, their blood type is AB. People whose cells do not carry any antigens have type O blood. The genes in a pair that determine a person's blood type can exist in any of three forms – A, B, or O. Blood type can be further broken down into Rh positive and Rh negative, depending on the presence of other antigens.

How many possible blood types are there?
Each member of a pair of genes that determines your blood type may be either A, B, or O. A and B are equally dominant. As a result, they are exhibited together as type AB. Because the O gene is recessive, you need two copies of it to be blood type O. The gene combinations for the four blood types are shown at right.

Inheritance of blood type
You inherit the genes that determine your blood type from your parents, one gene from your father and one from your mother. For example, if your mother has the A and O genes (making her blood type A) and your father has two O genes (making him blood type O), you have a 50 percent chance of inheriting one A and one O gene, making you type A. You also have a 50 percent chance of inheriting two O genes, making you type O (see right).

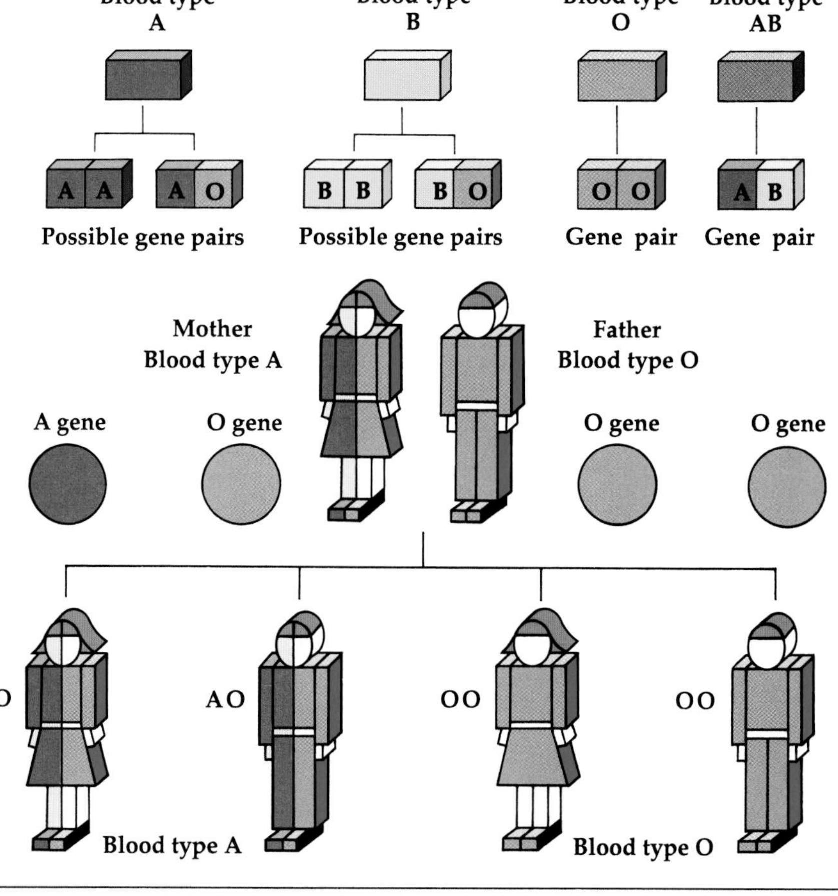

PREDICTING BLOOD TYPE

Because blood type is determined by genes, the possible blood types of children (center row) can be predicted if the blood type of both parents is known. The occurrence of some blood types among these children can also be ruled out (bottom row). Ruling out blood type has been used in paternity testing to exclude a man as the possible father of a child.

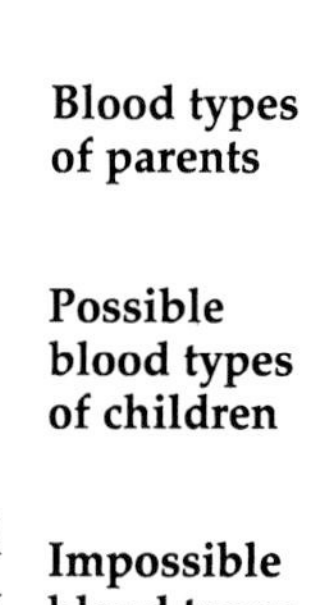

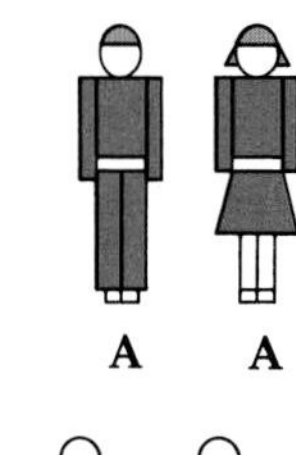
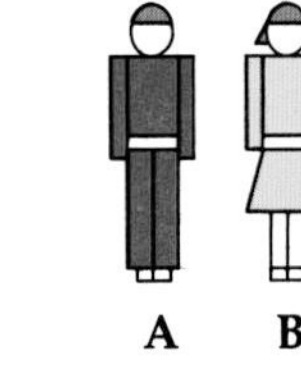
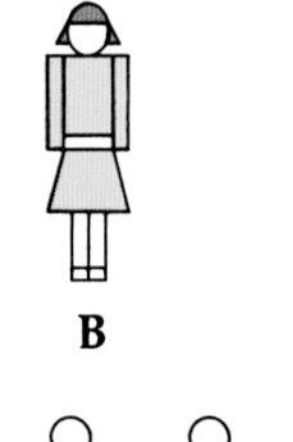

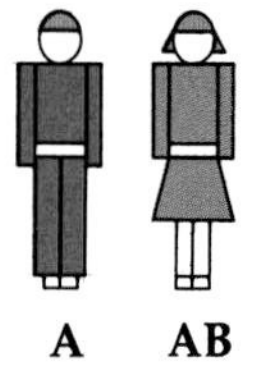
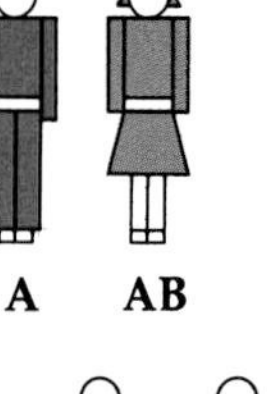
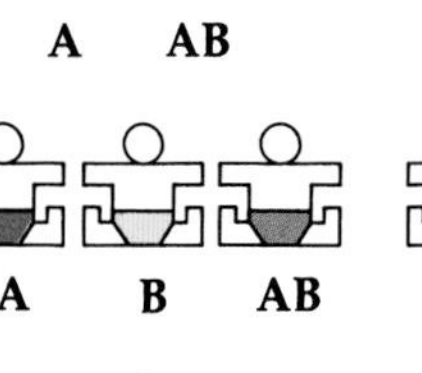

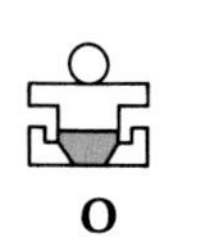

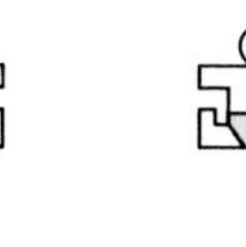

BLOOD TRANSFUSIONS

If you receive a transfusion of blood of a blood type different from your own, your immune system may recognize antigens (protein markers) on the surface of the donor red blood cells as foreign and launch an attack against them. This defensive response provokes a severe reaction. People with type O blood are considered universal donors because their red blood cells do not carry either A or B antigens. As a result, type O blood seldom produces an immune response in a recipient. People who have type AB blood are considered universal recipients because their red blood cells carry both the A and B antigens. Therefore, the presence of these antigens on donor blood of any type is not regarded as foreign and produces no immune response.

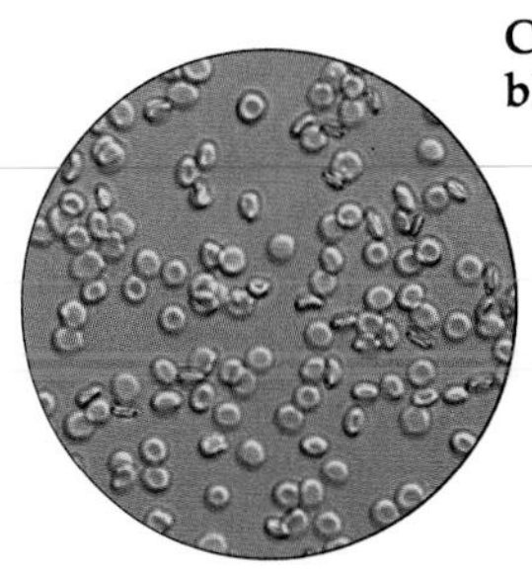

Compatible blood

Crossmatching blood
To detect incompatibility between donor and recipient blood, crossmatching is done before a blood transfusion. Doctors perform the test by adding a sample of donor blood to a sample of recipient blood. Blood that is incompatible looks clotted when viewed under a microscope.

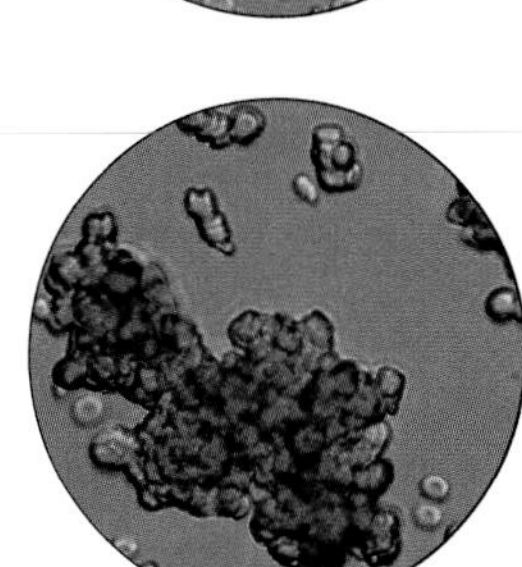

Incompatible blood

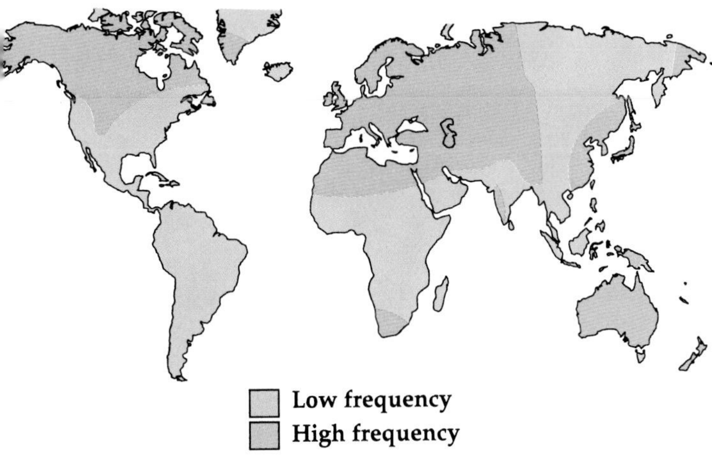

Distribution of blood type A
The frequency of blood types varies among geographically separate populations. The map at left shows the distribution of type A blood among various native populations. Scientists think the variation is caused by migration. If members of an immigrant group with different blood type patterns have children with members of the native population, the frequency of certain blood types in the native population may change.

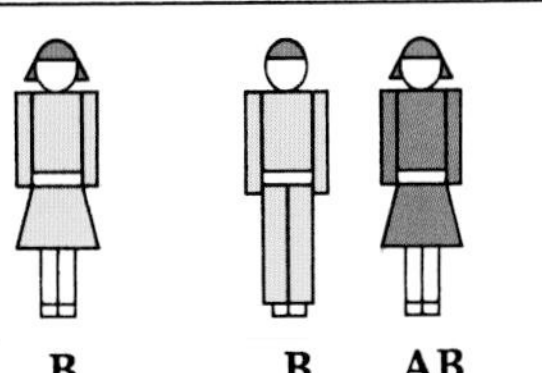
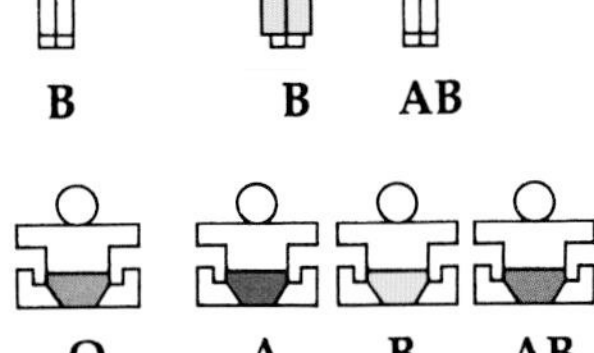
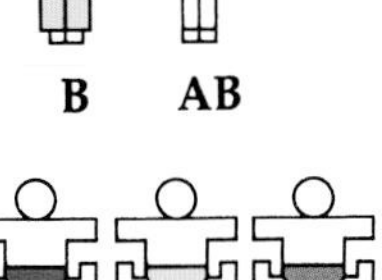
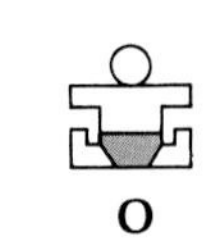
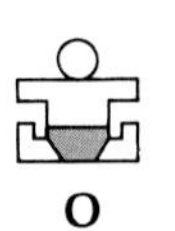

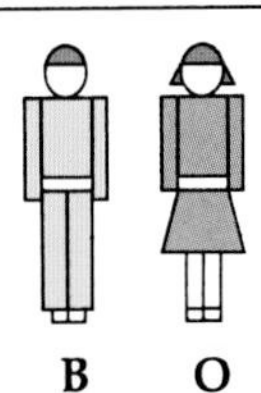
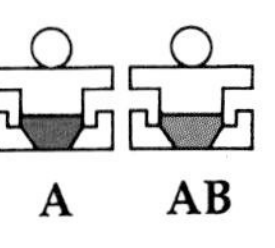
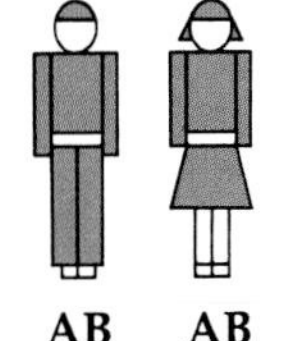

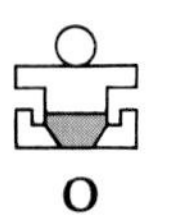

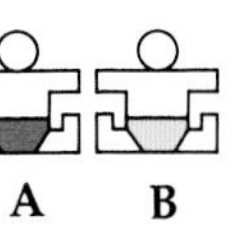
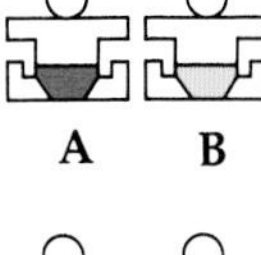
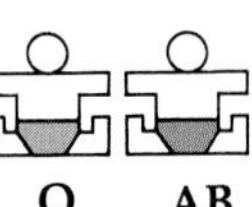
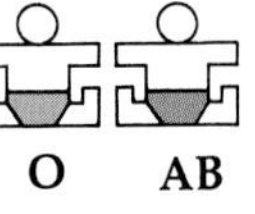

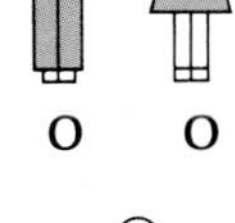
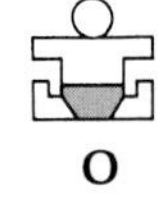
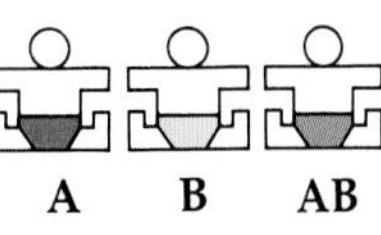
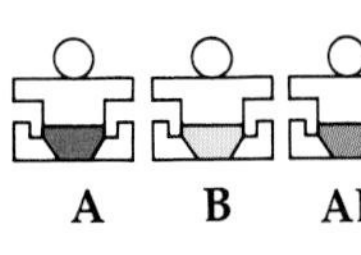

PATERNITY TESTING

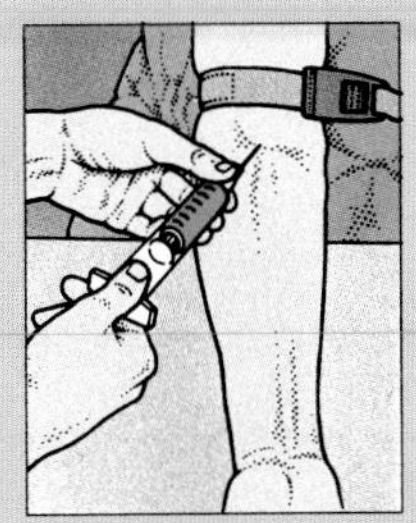

An unintended but useful by-product of the classification of blood type is its application to cases of disputed paternity. For example, if a child has type O blood, he or she must have inherited a copy of the O blood type gene from each parent. Therefore, if the child's mother is blood type A, she must have one O and one A gene. The father must also have an O gene and, therefore, cannot be of blood type AB. Although such blood tests cannot prove definitively that a man is the father, they can rule out a person if neither he nor the child's mother has the child's blood type. The technique of DNA fingerprinting (see page 113) is now replacing blood-type analysis in paternity testing because it offers a much higher probability of linking the biological father to a child.

SEX-LINKED INHERITANCE

The inheritance of characteristics that are determined by genes located on the sex chromosomes is called sex-linked inheritance. Sex-linked traits usually affect one sex more than the other.

Females have two X sex chromosomes, one inherited from each parent. Males have an X sex chromosome that comes from their mother and a Y sex chromosome from their father. Almost all sex-linked traits are governed by genes located on the X chromosome and are called X-linked traits. Other than male characteristics, the only trait thought to be dictated by genes located on the Y chromosome is the unusual one of hairy ears.

X-linked recessive inheritance

Most X-linked traits and disorders are recessive. Because females have two X chromosomes, the effects of a recessive X-linked gene are usually neutralized by the effects of a corresponding dominant gene on the other X chromosome. In males, however, the recessive gene produces an effect because males have only one X chromosome – the Y chromosome does not carry a corresponding dominant copy of the gene to neutralize its effect. For this reason, X-linked recessive traits affect mostly males. A male almost always inherits an X-linked trait or disorder from his mother, who is a carrier of the recessive gene but is herself usually unaffected. Color-blindness is an X-linked trait (see right). A number of serious disorders, including hemophilia and Duchenne type muscular dystrophy, also follow an X-linked recessive pattern of inheritance (see page 80).

In unusual circumstances, an X-linked recessive trait or disorder may affect a female. For example, a girl whose father is color-blind and whose mother is an unaffected carrier of the gene for color-blindness may inherit an X chromosome with the abnormal gene from both parents – and she will be color-blind.

COLOR-BLINDNESS

A person who is color-blind has difficulty seeing the difference between reds and greens. Like most X-linked recessive traits, color-blindness is more common in men than in women. When a man who is color-blind and a woman with normal genes for color vision have children, none of their children will be color-blind, but all of their daughters will be carriers of the abnormal gene. When a woman carrier and a man with normal color vision have children, each of their sons has a 50 percent chance of being color-blind and each daughter has a 50 percent chance of being a carrier.

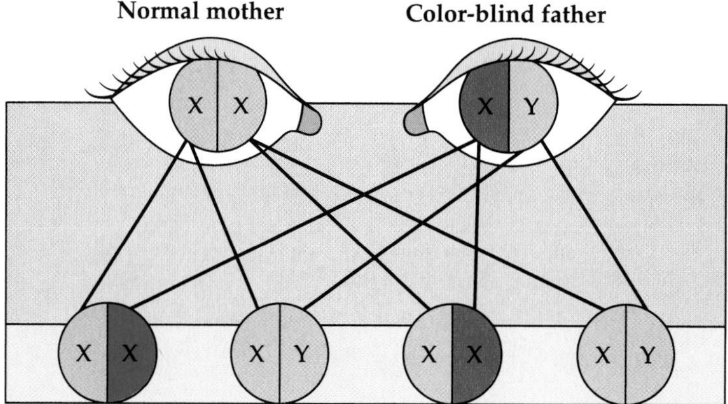

Color-blind father
If a man who is color-blind has children with a woman who has normal genes for color vision, all of their daughters will receive the affected X chromosome from their father. But, because they will also receive a normal X chromosome from their mother, they will be unaffected carriers of the gene. Because boys receive only the Y chromosome from their color-blind father and an X chromosome from their normal mother, they are not affected.

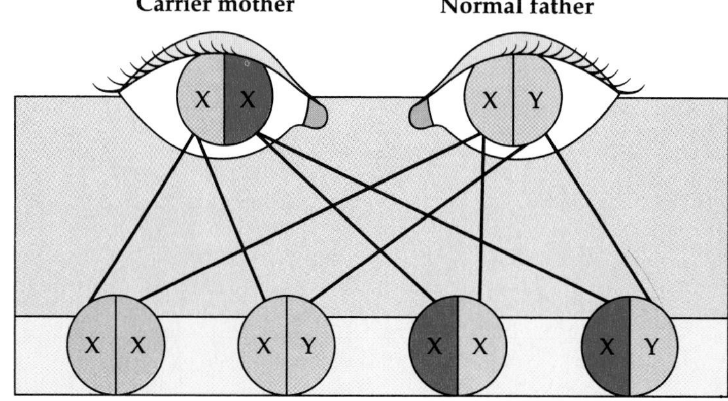

Carrier mother
If a female carrier and a man with normal vision have children, each of their daughters has a 50 percent chance of receiving the mother's affected X chromosome and of also being a carrier. Each son also has a 50 percent chance of receiving her affected chromosome but, because its effect is not neutralized by a normal X chromosome, he will be color-blind.

X-linked dominant inheritance

An X-linked dominant trait is one that is determined by a dominant gene on an X chromosome. X-linked dominant traits are rare – one of the few examples is a type of tooth discoloration.

Each child of a woman with an X-linked dominant disease gene has a 50 percent chance of inheriting her affected X chromosome and having the disease. Only the daughters of an affected man will inherit his X chromosome. This father-daughter inheritance pattern distinguishes an X-linked dominant trait from other dominant traits.

Y-linked inheritance

The way in which genes located on the Y chromosome are inherited is simple. Because only males have a Y chromosome, fathers transmit their Y chromosome only to their sons. Therefore, a man with a trait regulated by a gene on the Y chromosome will pass the trait on to all of his sons but to none of his daughters, because they inherit only his X chromosome. Apart from providing for maleness, little is actually known of the Y chromosome's role in heredity.

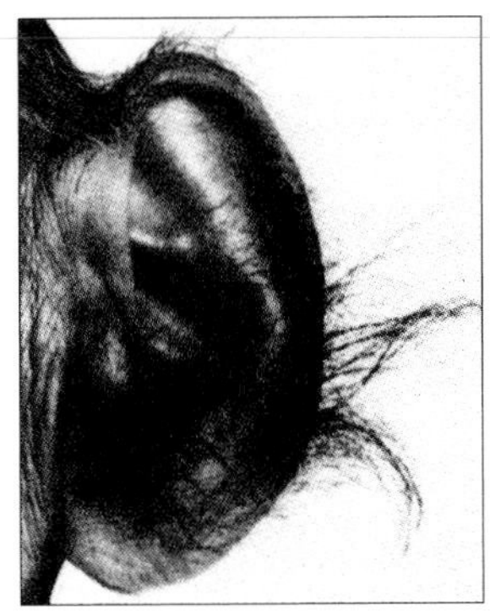

Y-linked characteristics
Other than maleness, the only trait that is thought to be Y-linked is hairy ears. Men with the Y-linked gene have long, stiff hairs in their ears, as shown at left. The trait is transmitted from father to son.

SEX-LIMITED TRAITS

Sex-limited traits affect structures or processes that exist in only one sex. Even though the genes responsible for these traits are not carried on the sex chromosomes, they almost always occur in one sex only. For example, baldness is a condition that appears to be governed by a single dominant gene located on one of the nonsex chromosomes. Because the level of activity of the gene and the degree of baldness are influenced by the presence of male sex hormones, the trait is limited to males.

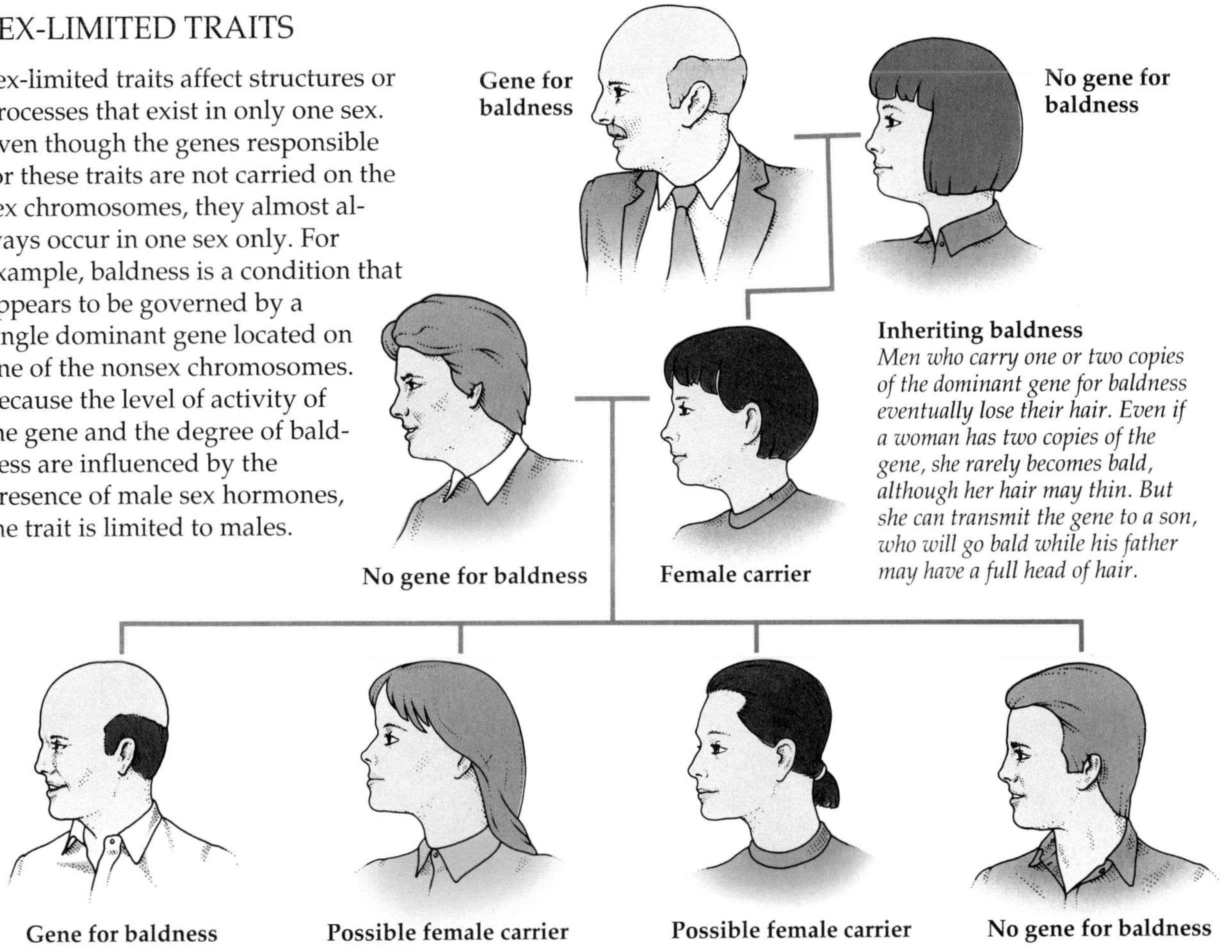

Gene for baldness

No gene for baldness

No gene for baldness

Female carrier

Inheriting baldness
Men who carry one or two copies of the dominant gene for baldness eventually lose their hair. Even if a woman has two copies of the gene, she rarely becomes bald, although her hair may thin. But she can transmit the gene to a son, who will go bald while his father may have a full head of hair.

Gene for baldness | **Possible female carrier** | **Possible female carrier** | **No gene for baldness**

COMPLEX INHERITANCE

The inheritance pattern of some characteristics, such as height and skin color, is difficult to determine because they are thought to result from the interaction of many different genes and environmental factors. For some of these traits, genes are the primary influence and environment plays a more minor role; for other traits, the relative importance of these influences may be reversed.

Skin color

Skin color is one of the best examples of a trait that is governed by many factors. There are at least 36 shades of skin color, ranging from almost black to almost white. Geneticists believe that variation in skin color among populations throughout the world is the result of the interaction of at least four genes. The intensity of skin pigmentation tends to be greatest in people who come from or live in areas close to the equator, while people who live farther from the equator tend to have lighter skin. This genetically determined variation in skin tone is probably the result of evolutionary adaptation to varying degrees of sunlight. But environment can also have an effect on the color and texture of your skin. For example, exposure to the sun makes your skin turn darker and, after prolonged exposure, become dry and wrinkled.

Height

Your height is also thought to be regulated by several genes. It appears that some genes ("tall" genes) strongly encourage growth, while other genes ("short" genes) encourage it less. The height you reach depends on the ratio of tall to short genes that you received from your parents. It is unlikely that a child will inherit all of one type of gene from both parents and, therefore, be exceptionally tall or exceptionally short. A child usually inherits a mixture of tall and short genes and reaches an average height. However, a child of tall parents tends to inherit more tall genes than short genes. Environmental influences can also affect a person's growth. Eating a poor diet, for example, can prevent a person from reaching his or her maximum potential height.

IMPRINTED GENES

In the mid 1980s, geneticists discovered that some genes, not just those carried on the X or Y sex chromosomes, have different effects on people, depending on whether the genes come from their mother or their father. These are known as imprinted genes. Scientists believe that there may be as many as 35 genetic disorders – including insulin-dependent diabetes, epilepsy, congenital heart disease, early-onset Huntington's chorea, and some childhood cancers – that are caused by imprinted genes. The full significance of this new discovery is still being evaluated.

Distribution of height
Height is thought to be determined by several genes. In most groups of people, individuals range from very short to very tall, with the majority of people falling in the middle. The average height of one population can differ dramatically from that of another, as shown in the graphs at right. The graphs compare the frequency of different heights for Pygmies from Zaire, Americans, and Dinkas from Sudan.

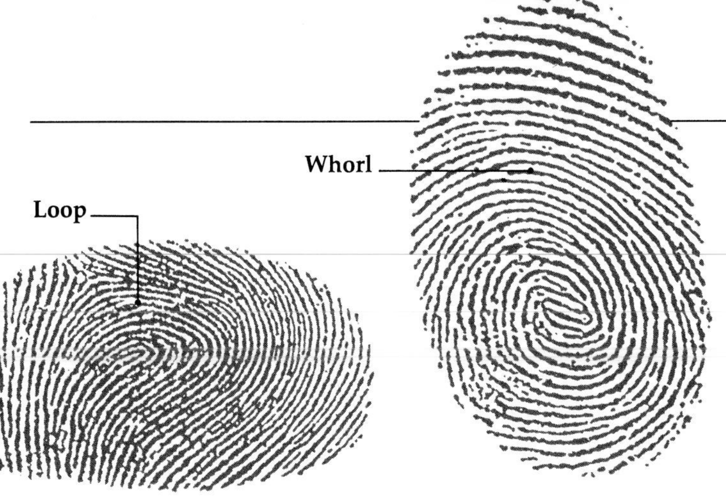

Fingerprints: no two are alike
Fingerprint pattern, which is a trait influenced by several genes, has been studied extensively. The three basic fingerprint patterns – the arch, the loop, and the whorl – are shown here. No two people – not even identical twins – have identical fingerprints. Fingerprint patterns are determined by an unknown number of genes that interact with the environment inside the uterus during fetal development.

Body build
Body build is an example of a trait of complex inheritance that probably represents evolutionary adaptation to the environment. Although there are exceptions to the rule, many differences in body build among populations of the world are thought to be influenced by the climate in which they live. For example, Eskimos tend to be stout, which helps them conserve body heat (right). Conversely, some Africans are tall and thin, which provides them with more body surface to help them dissipate heat (above right).

ASK YOUR DOCTOR HEREDITY

Q I recently read that you can inherit a predisposition to alcoholism. Is this true?

A Researchers are gathering an increasing amount of evidence indicating that genes play an important role in susceptibility to alcoholism. However, a person with the genes for alcoholism will obviously not become an alcoholic unless he or she drinks. Most geneticists believe that alcoholism results from a complex interaction of one or more genes and environmental factors.

Q My parents are both extroverted, strong-willed people, and so am I. Does that mean I inherited my personality from them?

A Like other complex characteristics, your personality is regulated by the interplay of many of your genes and the influences of your environment. Because the inheritance of personality does not follow a simple pattern, it is not easy to determine. Although genes are thought to play a part, many geneticists believe that personality is influenced much more by the way people are raised and their life experiences.

Q My father and his father both had full heads of hair into old age, but I am becoming bald at 30. If baldness is inherited, from whom did I inherit it?

A You probably inherited the gene for baldness from your mother. Women can carry the gene without any noticeable effect because their hormones protect them from extreme hair loss. If your father had the gene, he would most likely have been bald because the male sex hormones make the gene active.

CHAPTER THREE

GENES AND DISEASE

Many of the chronic diseases that affect people today are determined, at least in part, by genes. It is difficult to know the true incidence of diseases that are influenced by genetic factors because many disorders do not appear until late in a person's life. By age 25, nearly 5 percent of the population are affected by disorders in which genes play a large part. Up to 60 percent are affected by such disorders at some time in their lives. About one in 30 babies has a birth defect, a chromosome abnormality, or a disorder caused by a single gene. The incidence of genetic mistakes is much higher at conception, but most of the severely defective fertilized eggs are miscarried. Genetic disorders contribute significantly to illness and death, particularly during childhood, but many of these disorders are now being treated successfully.

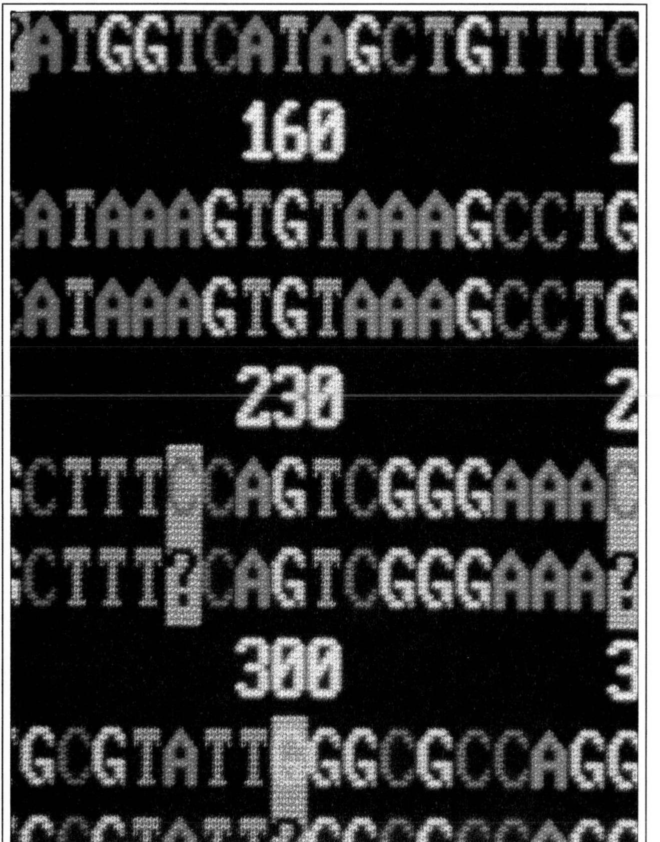

There are three main categories of genetic diseases – chromosome abnormalities, single-gene disorders, and disorders that result from an interaction of both genes and environment. Chromosome disorders, such as Down's syndrome, are caused by abnormalities in the number or structure of chromosomes. A few chromosome abnormalities are inherited, but the majority occur as random events at the time of conception. The risk that these sporadic genetic mistakes will happen twice in a family is usually low. Single-gene disorders are caused by a single defective gene or pair of genes. Single-gene disorders are usually inherited and the risk of recurrence in a family is often high. Defects in individual genes are sometimes the result of new mutations that occur before conception at the time an egg or sperm is formed. Sickle cell anemia, cystic fibrosis, and muscular dystrophy are among the most common and serious single-gene disorders.

Some disorders occur in people who are susceptible as a result of the interaction of several defective genes and environmental factors such as pollution and cigarette smoke. Although the inheritance of these disorders is not clear-cut, they tend to occur more frequently in relatives of affected people than in the general population. Many birth defects (such as clubfoot, spina bifida, and congenital heart disease) and many common diseases that become apparent later in life (such as diabetes and coronary heart disease) result from a combination of interrelated genetic and environmental factors.

The final section of this chapter describes the role that genes play in cancer. Research indicates that some people may inherit genes that increase their risk of cancer.

HAZARDS TO A FETUS

Exposure to the following agents can harm a fetus.

- ◆ Alcohol can stunt fetal growth and cause mental retardation.
- ◆ Tobacco smoke can stunt fetal growth and cause miscarriage or premature birth.
- ◆ Some drugs, such as retinoid drugs for acne, can cause severe organ abnormalities.
- ◆ Some viruses infecting a pregnant woman can cause fetal brain and heart damage, deafness, and blindness.

Syndromes

Some malformations occur together more often than would be expected by chance. These multiple malformations, which are called syndromes, are not as common as single malformations and may have many causes, both environmental and genetic, or may have no known cause. They are among the most serious congenital abnormalities.

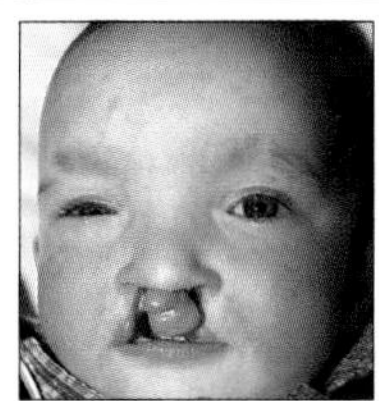

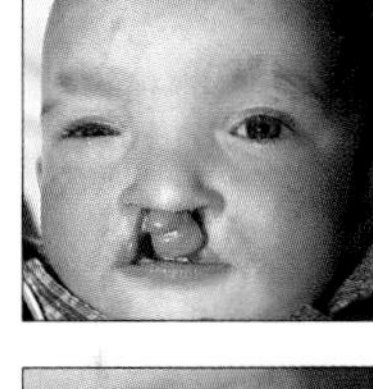

Cleft lip and palate
Cleft lip is a vertical split in the upper lip that may extend up to the nose (above left) and can occur with cleft palate (a gap in the roof of the mouth). The child in the photograph below left has had reconstructive surgery, which can almost completely correct these abnormalities.

PYLORIC STENOSIS

In a baby born with pyloric stenosis the circular muscle at the outlet of the stomach is thickened, obstructing the passage of food into the intestine. The disorder, which mostly affects male babies, causes severe vomiting about 3 to 4 weeks after birth. The estimated risks to family members are illustrated below.

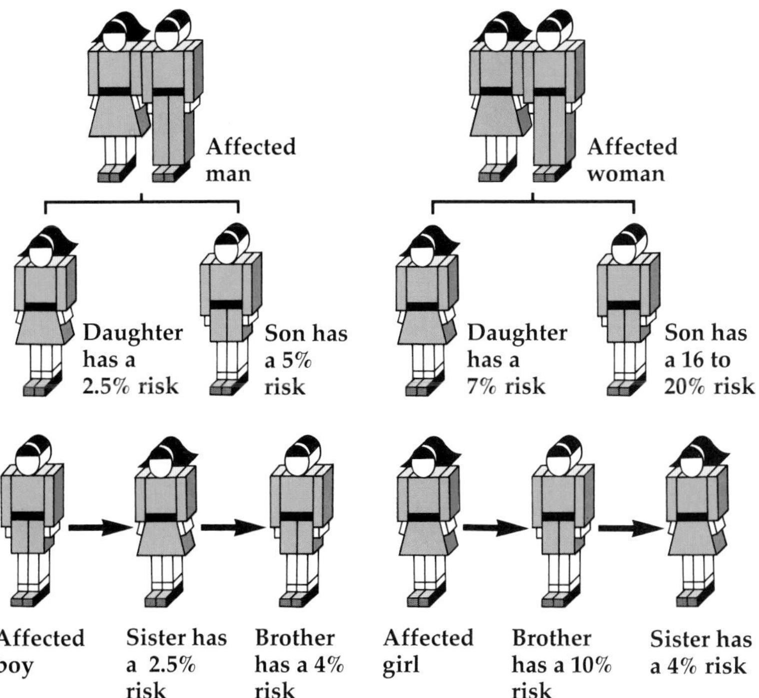

ASK YOUR DOCTOR

CONGENITAL ABNORMALITIES

Q My son was born with a heart defect. Did this happen because I was so shocked by my father's death during my pregnancy?

A No. There is no evidence that emotional trauma during pregnancy can cause a congenital abnormality in a child.

Q My first child was born with a diaphragmatic hernia and lived for only 48 hours. My doctor says that the chance of this happening again is small. Are there any tests I can have during pregnancy?

A Yes. Although your risk of having another affected child is low, an ultrasound scan during pregnancy can detect diaphragmatic hernia (in which abdominal organs protrude through the diaphragm into the chest). In addition to eliminating the possibility of this problem, a normal scan will reassure you that the fetus is unlikely to have any other major structural defects.

Q I was born with a cleft lip and palate. Will my children also have these abnormalities?

A Not necessarily. Your children's risk depends on a number of factors. If you have no other congenital abnormalities, and none of your relatives has the defect, the risk to each of your children is about 1 in 25. The risk is higher if other relatives are affected. Cleft lip and palate may be caused by drugs taken during pregnancy, but, in these cases, other abnormalities are usually present as well. If drugs caused your condition, your child's risk is almost the same as that of the general population.

GENES AND CANCER

CHANGES OR MUTATIONS in the DNA of cells can lead to cancer. Some of these mutations are inherited directly, but most are caused by by-products of the body's chemistry or environmental toxins. It is now thought that, out of the approximately 100,000 genes in each of your cells, more than 20 specific genes can lead, as a result of mutation, to the development of cancer.

Cancer is the unregulated division of cells. Although most types of cancer are not actually hereditary, a person can inherit a susceptibility to some cancers.

CANCER-CAUSING GENES

Two types of genes are thought to be involved in the development of cancer: proto-oncogenes and antioncogenes. Proto-oncogenes promote the normal growth and division of cells. Most of the time these genes are inactive, or switched off. But a mutation in a proto-oncogene can permanently switch it on, turning it into an oncogene. Oncogenes, which operate like gas pedals that are stuck, lead to the excessive cell division that is characteristic of cancer. Antioncogenes normally serve as brakes to keep cell division in check. But when a mutation inactivates an antioncogene, it no longer performs its job of preventing runaway cell division and cancer results.

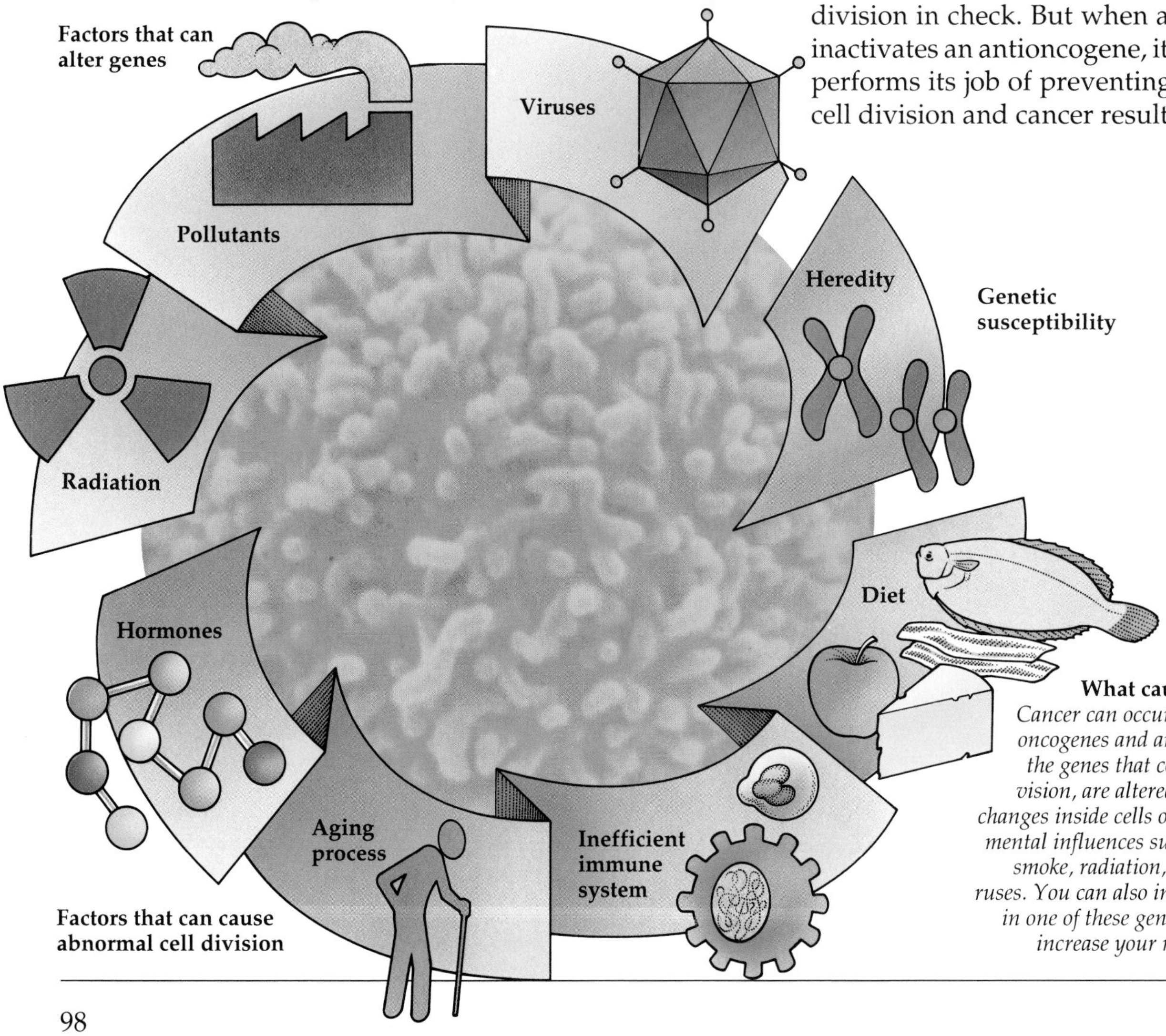

Factors that can alter genes

Pollutants

Viruses

Heredity

Genetic susceptibility

Radiation

Hormones

Diet

Aging process

Inefficient immune system

Factors that can cause abnormal cell division

What causes cancer?
Cancer can occur when proto-oncogenes and antioncogenes, the genes that control cell division, are altered by chemical changes inside cells or by environmental influences such as tobacco smoke, radiation, and some viruses. You can also inherit a defect in one of these genes, which can increase your risk of cancer.

CANCER AND HEREDITY

If you have a relative with cancer or one who died of cancer, you are not necessarily at increased risk of developing it yourself. Some types of cancer, such as cancer of the cervix, lung, or bladder, are rarely inherited. Nevertheless, some families appear to have a susceptibility to a certain cancer. Between 4 and 10 percent of all cases of cancer of the breast, colon, and ovary are thought to have a genetic component. At-risk families cannot be identified until a particular cancer has occurred more than once in the family and in a person of younger-than-average age for that cancer.

Genetic susceptibility to cancer

Usually, cancer develops only after a proto-oncogene or an antioncogene has undergone one or more changes. A person who inherits an already altered gene has an increased risk of cancer. When these altered genes are exposed to certain internal or external influences, such as hormones inside the body or radiation from sunlight, cancer may develop.

In some cancers, inherited mutations make DNA unable to repair itself after it has been damaged. Efficient repair of genetic damage protects the body from cancer. Because people vary in their ability to repair DNA, their risk of developing some cancers depends on how efficient their repair mechanisms are.

SINGLE-GENE DISORDERS AND CANCER

A few single-gene disorders – including neurofibromatosis, familial polyposis, and tuberous sclerosis – make people susceptible to some types of cancer. These disorders usually follow a pattern of autosomal dominant inheritance, which

BREAST CANCER

Geneticists have identified a gene on chromosome 17 that causes an inherited form of breast cancer in women under 40. Women with a family history of early breast cancer have a more than 30 percent chance of developing it by age 40, compared with a less than 1 percent risk for women with no family history.

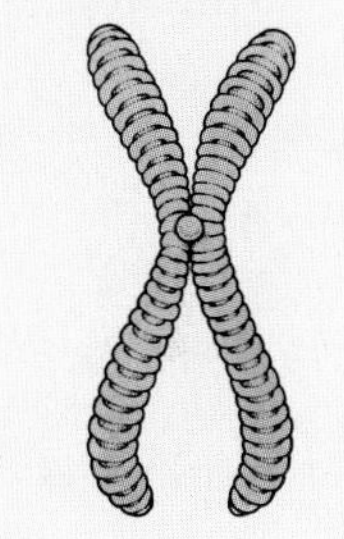

Ovarian cancer

As shown in the family inheritance pattern below, ovarian cancer sometimes affects several women in a family and probably results from an abnormal dominant gene. In these families, the daughters of an affected woman have up to a 50 percent chance of also developing ovarian cancer. Researchers are working to identify the cancer-causing gene, which promises to improve the ability to predict a woman's risk.

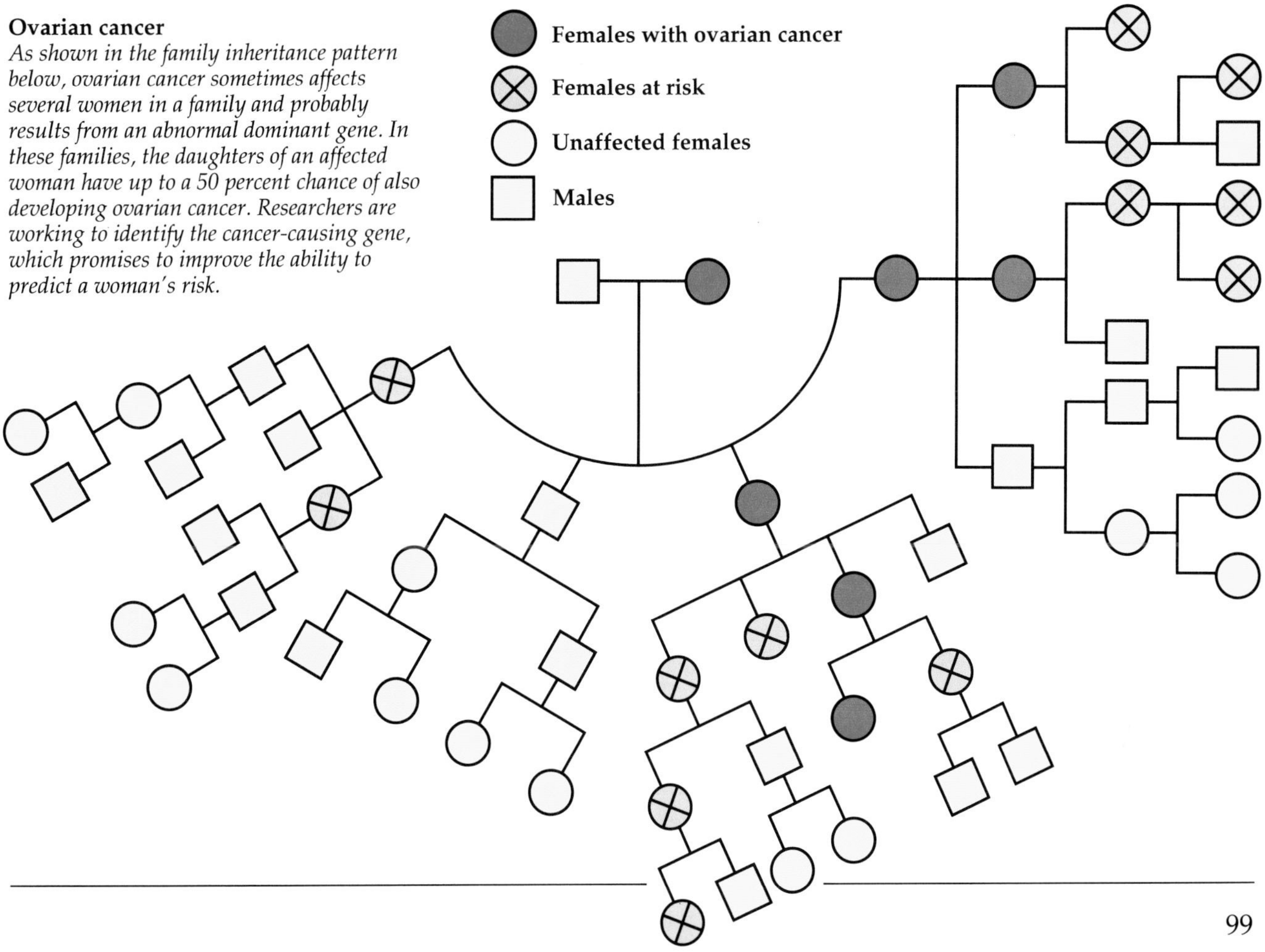

Females with ovarian cancer
Females at risk
Unaffected females
Males

means that only one gene is necessary to cause the disorder. Each child of a person with the abnormal gene has a 50-50 chance of inheriting it, and, as a result, of being susceptible to the cancer.

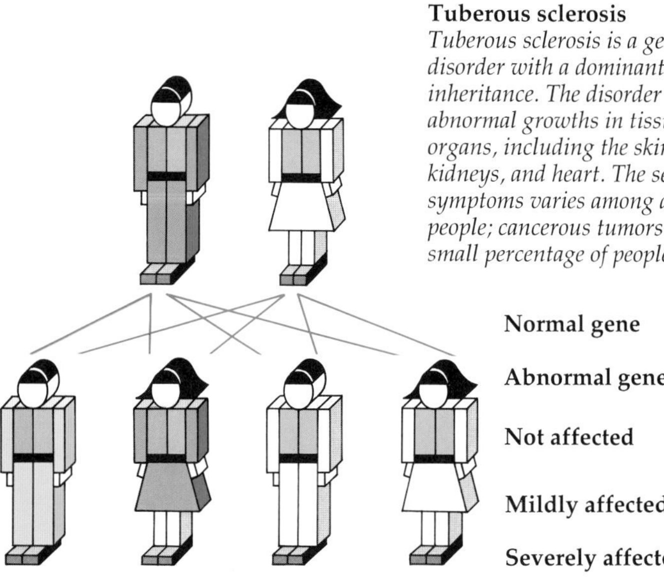

Tuberous sclerosis
Tuberous sclerosis is a genetic disorder with a dominant pattern of inheritance. The disorder can cause abnormal growths in tissues and organs, including the skin, brain, kidneys, and heart. The severity of symptoms varies among affected people; cancerous tumors occur in a small percentage of people.

Neurofibromatosis

Approximately one in every 3,000 people has an abnormality in the gene that regulates the division of nerve cells. The disorder that can result, called neurofibromatosis, produces patches of brownish pigment on the skin and noncancerous tumors that appear as bumps on the skin. One of these benign tumors may develop into cancer and spread. Occasionally, tumors or cancerous growths form on a person's brain and spinal cord. The effects of neurofibromatosis vary from hardly noticeable to severe; nine out of 10 affected people do not develop cancer.

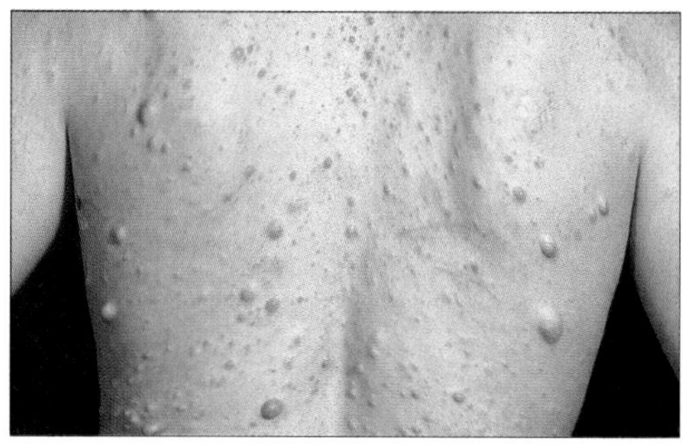

Signs of neurofibromatosis
A person with neurofibromatosis has patches of pigmented skin and may develop numerous soft, fibrous swellings that grow on the nerves in the skin.

Familial polyposis

About one person in 10,000 has familial polyposis, an inherited disorder characterized by multiple abnormal growths called polyps inside the intestines (mainly the large intestine). This disorder is caused by an altered gene that controls the division of cells lining the intestines. People with familial polyposis tend to develop the polyps during their late teens and early 20s. By age 30, they may have thousands of polyps. Because these cells have already undergone the genetic changes that can lead to cancer, the risk that the polyps will become malignant is high. Without preventive treatment, such as surgical removal of the polyps or even removal of the colon itself, a person with the disorder usually develops colon cancer by the age of 40.

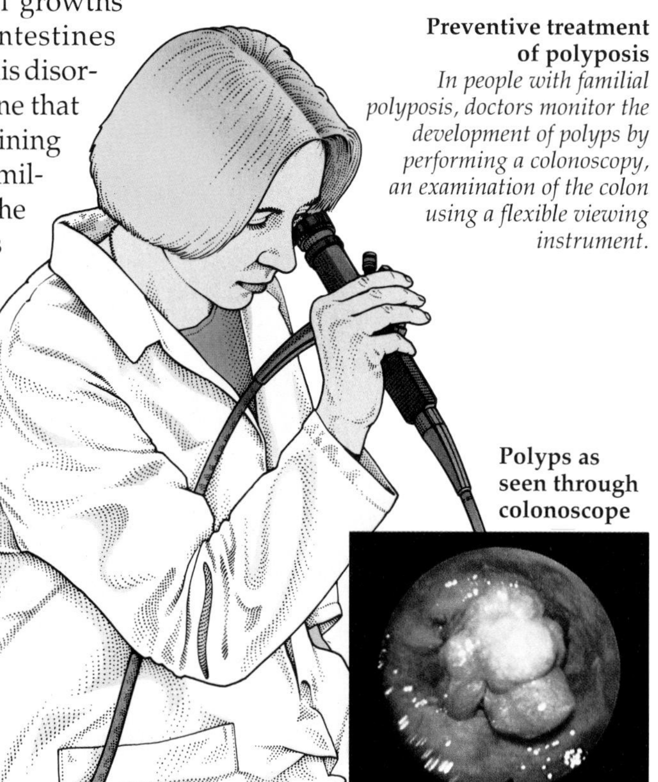

Preventive treatment of polyposis
In people with familial polyposis, doctors monitor the development of polyps by performing a colonoscopy, an examination of the colon using a flexible viewing instrument.

Polyps as seen through colonoscope

CHROMOSOMES AND CANCER

Chromosome defects are characteristic of some forms of cancer. For example, about 90 percent of people who develop chronic myeloid leukemia carry a chromosome rearrangement known as the Philadelphia chromosome. The abnormal chromosome consists of chromosome 9 with a small piece of chromosome 22 attached. The defect tends to switch on a cancer-causing gene (an oncogene), which results in this type of leukemia.

Retinoblastoma in one eye
Retinoblastoma appears initially (often in childhood) as a whiteness in the eye. About 90 percent of small tumors that occur in only one eye can be treated effectively with either surgery or radiation therapy.

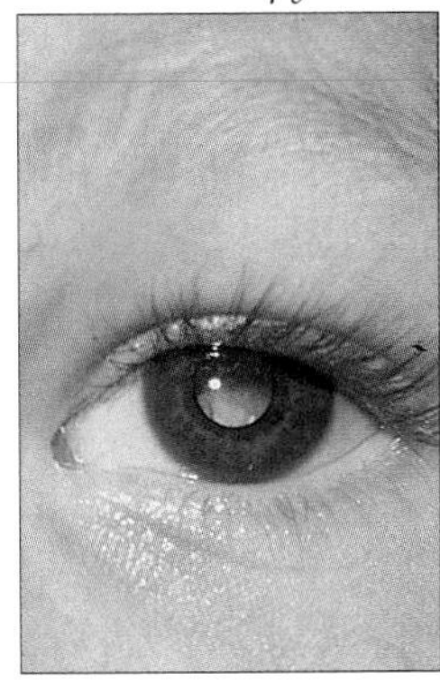

Retinoblastoma

Cancers that appear early in childhood are more likely to have been inherited than cancers that appear later in life. About half of all cases of retinoblastoma, which is cancer of the retina (the light-sensitive layer at the back of the eye), are inherited as the result of an altered anti-oncogene (one of the "brakes" that halt cell division) found on chromosome 13. In about one fourth of cases, children are born with a tumor in each eye. Brothers and sisters of a child with retinoblastoma in both eyes have a 40 percent risk of having it if one of their parents is also affected. Occasionally, a child with no family history develops only one eye tumor. The relatives of that child, including his or her future children, have a low risk of developing retinoblastoma.

DETECTING INHERITED TUMORS

Because the symptoms of neurofibromatosis and tuberous sclerosis are apparent on the surface of the body, people with these disorders are usually identified before age 10. Close family members who may also be at risk can then be examined for the disorder.

Carriers of some inherited diseases, such as familial polyposis, can be identified using DNA analysis. The DNA in blood samples taken from family members is examined for the disease-causing gene (see GENETIC ANALYSIS on page 108). If the gene is not found in a blood sample, that person probably needs no further examination and his or her children are at little risk of developing polyposis.

CANCER MONITORING FOR AT-RISK FAMILIES

Although a strong family history may implicate heredity as the cause of a cancer, this is not always the case. Even if the tendency to develop a cancer is inherited, it is usually not possible to identify the abnormal gene and, therefore, the members of the family who may be at risk. All members of an at-risk family should be examined regularly for the first signs of cancer. The earlier a cancer is detected, the more effective the treatment is likely to be. If you have a family history of a cancer, your doctor may recommend that you have regular tests to detect cancer at an early, treatable stage.

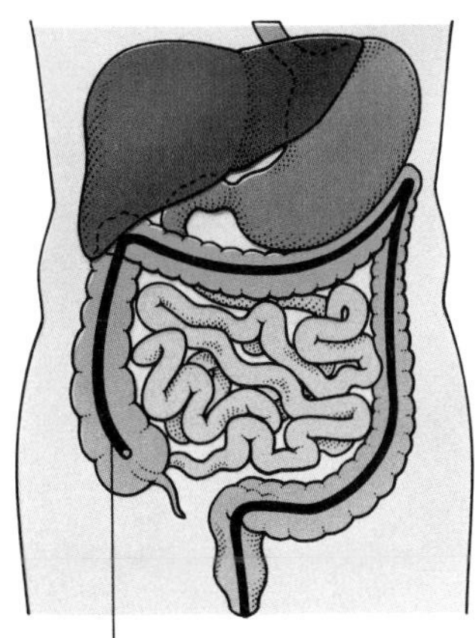

Examining the colon
If your family has a risk of colon or rectal cancer, your doctor may recommend that you have a colonoscopy (an examination of the entire colon to detect polyps) every year.

Examining the breasts
Women of any age with a family history of breast cancer should agree on a plan with their doctor to have frequent breast examinations and mammograms. They should also do a monthly self-examination of their breasts.

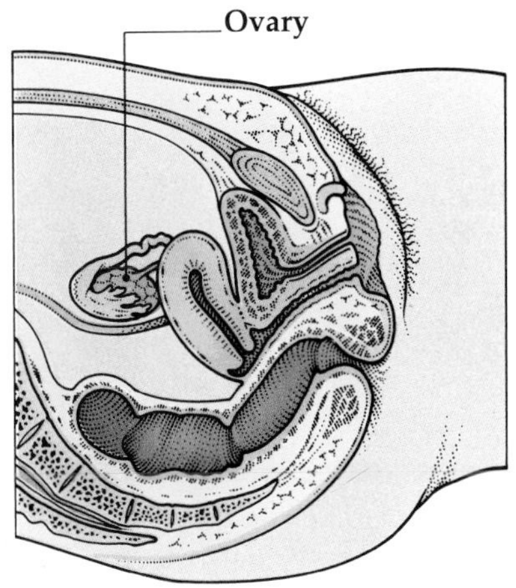

Examining the ovaries
To check for ovarian cancer, doctors use clinical examination and ultrasound (which is done to obtain an image of the ovaries). An ultrasound scan can detect ovarian cancer at an early stage.

CHAPTER FOUR

GENETIC COUNSELING, DIAGNOSIS, AND TREATMENT

An estimated 7 percent of children are born with a birth defect, mental retardation, or a genetic disorder. These defects are the major killer of children in the first year of life. Forty percent of defects present at birth are known to result from genes alone or from a combination of genetic and environmental factors. Although the causes of the remaining 60 percent have not yet been determined, many of these abnormalities may be influenced, at least in part, by genes. Some genetic disorders – such as Huntington's chorea and some forms of blindness and deafness – do not show up until later in life; this type of disorder affects one in 80 adults.

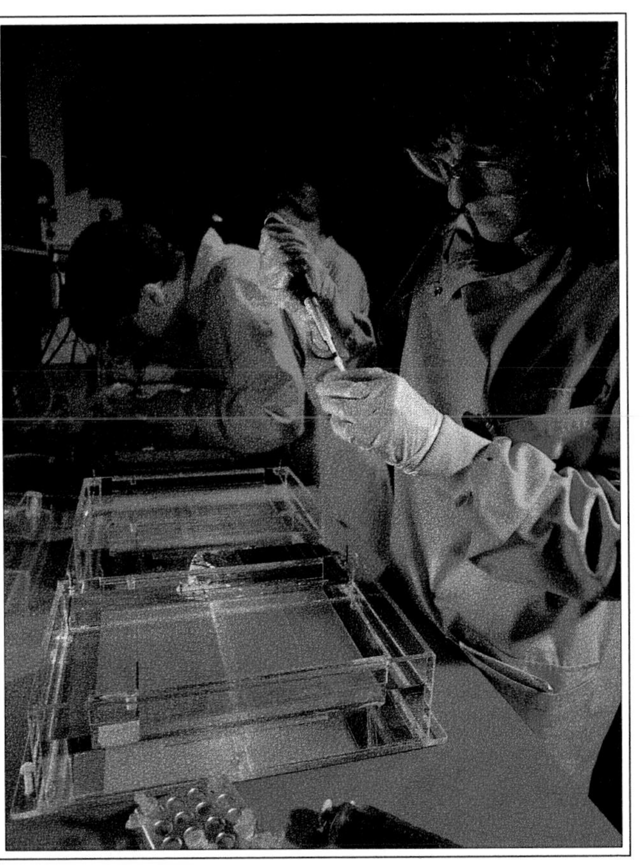

New knowledge about the genes that cause disease and the ways in which birth defects and genetic disorders occur has helped doctors make more accurate diagnoses. They can perform specific tests that allow them to analyze a person's genetic makeup, inform people about how they can avoid transmitting a genetic disorder to offspring, and provide treatment for many disorders.

When a couple has a child with a birth defect or an inherited disorder, the first step is to get a diagnosis. Many disorders have similar symptoms but very different causes. In many cases, it is possible to obtain a laboratory diagnosis. If a child's defect is shown to have a genetic cause, the parents may be referred to a genetic counselor. The counselor can provide them with the known facts about the disorder and offer them more tests, if necessary. The counselor also evaluates a couple's risk of having another child with the same disorder.

Genetic counseling also benefits people who are themselves affected by a genetic disorder or have a relative who is. They may want to know their risk of passing an abnormal gene on to children. Members of some ethnic groups are at increased risk for particular genetic disorders, such as blacks for sickle cell anemia and Ashkenazi Jews for Tay-Sachs disease. They may decide to be tested to find out whether they carry the disease-causing gene. Couples found to be at increased risk of having a child with a serious genetic disorder can choose from a variety of family planning options. Prenatal tests during pregnancy can detect a number of abnormalities in a fetus. If a test diagnoses a severe, untreatable disorder, a couple may choose to terminate the pregnancy. Prenatal detection can also help prepare parents psychologically for the birth of a child with a genetic defect and give hospital staff time to make preparations to care for an affected child immediately after birth.

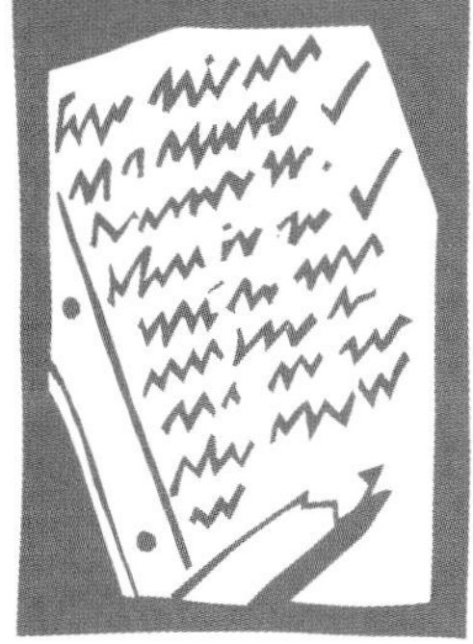

EVALUATING YOUR RISK

If you are at increased risk of having a child with a genetic disorder, tests can determine whether or not you are a carrier of the disease-causing gene. The test results are used by genetic counselors to evaluate the chances of your transmitting a defective gene to your children. If you are found to be at risk of having a child with a genetic disorder, you might choose an alternative route to parenthood, such as artificial insemination or adoption.

WHO CAN BENEFIT FROM MEDICAL ADVICE?

Couples with an affected child
Couples who have already had a child with a birth defect or a genetic or chromosome disorder.

Couples with a family history of disease
Couples in which one or both members have a genetic disorder or a birth defect or have close relatives with one.

Couples from some ethnic groups
Couples from specific ethnic groups who are at increased risk for certain genetic disorders.

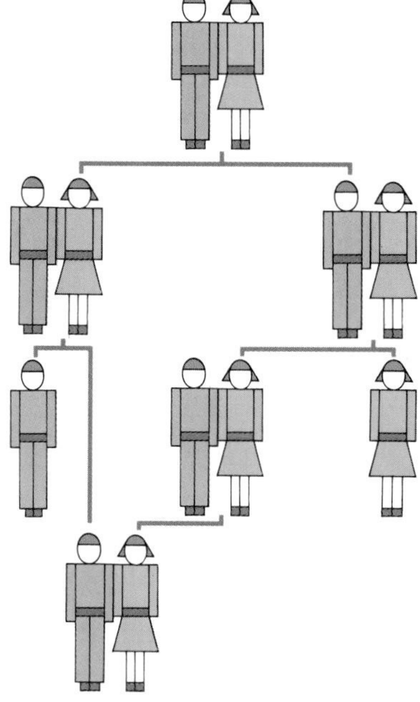

Related couples
Couples related by blood, such as a man married to his cousin's daughter.

At-risk women
Women who want to become pregnant and are 35 or older or who have had three or more miscarriages.

Doctors and genetic counselors can help individuals, couples, and families who are at risk of being affected by a genetic disorder by advising them, arranging for genetic testing, interpreting the test results, and providing treatment.

BENEFITS OF GENETIC COUNSELING

Couples who have had a child with a genetic disorder or a birth defect and are worried that a second child will also have one frequently seek genetic counseling. Such couples often believe that they are incapable of having any normal children or that they are much more likely than other couples to have another child with a birth defect or genetic disorder. These concerns are often unfounded. Not all genetic disorders are inherited – some result from new gene mutations that occur when a parent's egg or sperm is formed – and most birth defects are caused only partly by genes. In many cases, a genetic counselor can reassure concerned parents that the risk to their next child is significantly lower than they imagined.

Studying the family's health history and, in some cases, the results of blood tests of prospective parents and other family members can help genetic counselors evaluate the risk of a particular disorder. If the couple's risk is found to be increased, the counselor can explain to them the range of family planning options available (see opposite).

FAMILY PLANNING OPTIONS

All couples who are considering starting a family want to have a healthy child. Couples who are at increased risk of having a child with an abnormality may want to see a genetic counselor to discuss the various family planning options available to them.

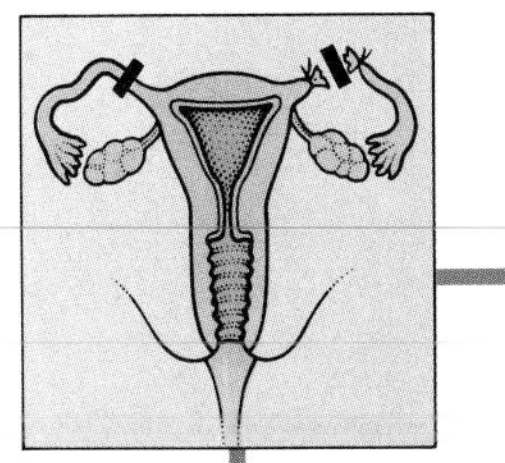

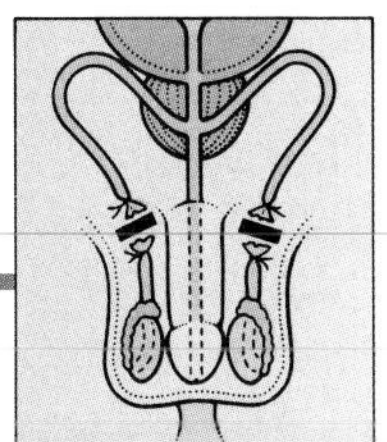

Sterilization and adoption
If a couple considers their risk of having a child with a genetic disorder to be unacceptable, one of them may choose to be sterilized. Couples who do this but who still want to have children may decide to adopt.

Determining the risk
The genetic counselor discusses with the couple their risk of having a child with a genetic disorder.

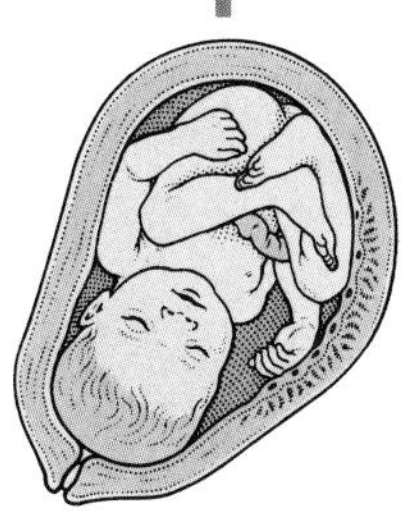

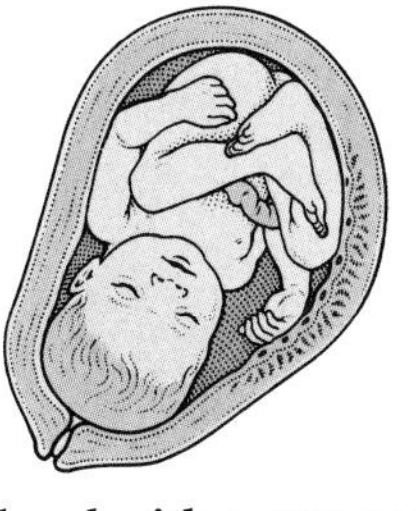

Going ahead with a pregnancy
Every couple has the right to give birth to a child, even if that child is at risk. Some couples do not wish to have prenatal tests while others choose to have some type of prenatal testing.

Contraception
Some couples postpone starting a family in the hope that advances in DNA testing, prenatal diagnosis, and treatment for a particular genetic disorder will increase and improve their options.

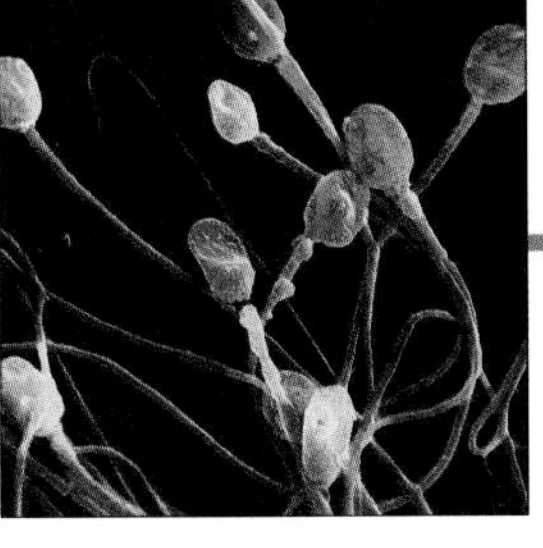

In vitro fertilization
If a woman carries a genetic abnormality, and she and her partner are determined to have a child, sperm provided by the man can be used to fertilize a donated egg in a test tube in the laboratory. A doctor then inserts the fertilized egg into the woman's uterus. In this way, a woman can give birth to a healthy baby whose genes come from her partner and a donor.

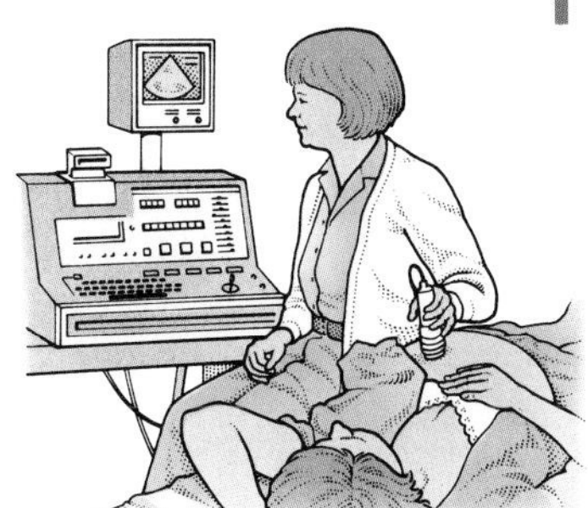

Prenatal testing
Some genetic disorders and birth defects can be detected by prenatal tests. For some abnormalities, arrangements can be made for the baby to be delivered in a hospital that specializes in treating birth defects. In other cases, a couple may choose to terminate the pregnancy.

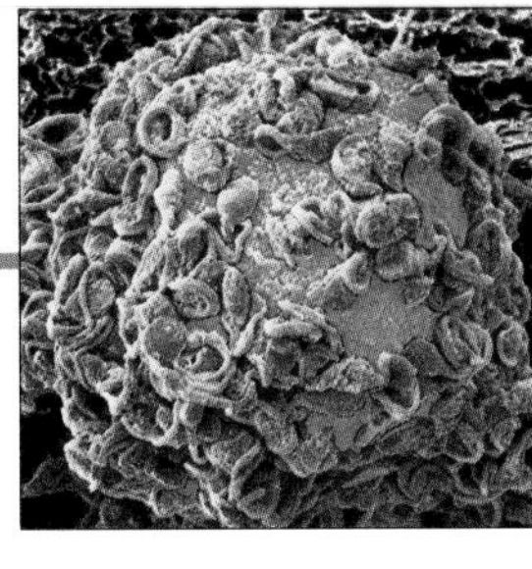

Artificial insemination
If a man has the dominant gene for a serious genetic disorder or if both members of a couple carry the recessive gene for a disorder, artificial insemination with donated sperm can be used to fertilize the woman's egg. In this technique, a doctor introduces sperm donated by a healthy man into the woman's uterus close to her time of ovulation.

BIRTH

Prenatal diagnosis

Prenatal testing (see PRENATAL TESTING AND DIAGNOSIS on page 122) often provides a couple with the reassurance that their baby will not be affected by a particular genetic disorder. However, such tests may also indicate an increased risk of an abnormality. When a serious abnormality is detected, the genetic counselor discusses with the couple the degree of mental and physical impairment that the child is likely to have; available treatment for the condition, including prenatal treatment such as surgery performed on the fetus inside the uterus; and any promising advances in treatment that may soon be available. Based on the advice they receive and their personal values, expectations, and circumstances, the couple decides on the course of action they will take. One option is abortion. Some couples would never consider abortion. However, they still may want to find out during pregnancy if there is an abnormality so that they have time to prepare themselves mentally and emotionally for their child's birth.

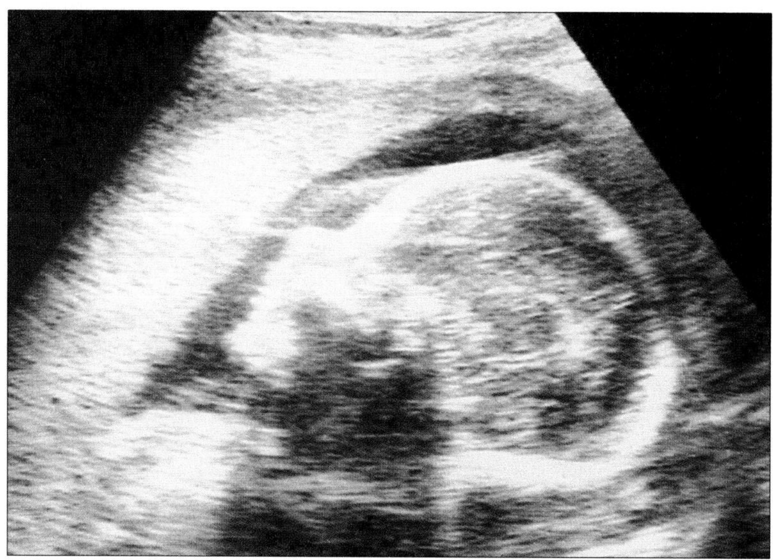

Ultrasound
The ultrasound scan above shows the head of a normal fetus. Doctors use ultrasound to monitor the development of a fetus and to diagnose skeletal abnormalities.

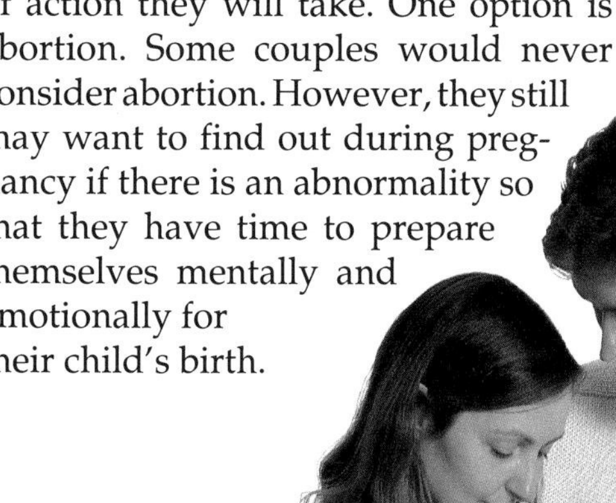

Personal choice
Some couples choose to decline any tests offered during pregnancy. Other couples want to know of an abnormality and have prenatal testing as early as possible to allow them to consider their options.

PRENATAL SCREENING

All pregnant women are offered a number of tests during pregnancy to detect abnormalities in a fetus. In most cases these tests reassure a couple that the pregnancy is progressing normally. A blood sample taken at about the 16th week of pregnancy can identify fetuses at increased risk of spine defects such as spina bifida (see page 95). Doctors use ultrasound scans to detect major physical abnormalities in a fetus.

Because of the increased risk of having a baby with Down's syndrome, pregnant women 35 and older are offered a test that analyzes a fetus's chromosomes.

SCREENING OF NEWBORNS

All newborns are screened for some inherited disorders for which early detection and treatment can improve the outlook. For example, inherited metabolic disorders such as phenylketonuria (PKU) can be identified with a simple blood test. Phenylketonuria causes mental retardation unless the affected child follows a restricted diet that is maintained throughout his or her life. All newborns in the US are tested for PKU and hypothyroidism (a disorder that can prevent normal growth and development). Some states also provide screening for metabolic disorders such as galactosemia (a disorder in which the body cannot digest a type of milk sugar).

Some genetic disorders, such as Duchenne type muscular dystrophy, can also be detected in newborns, but newborn screening for these disorders is not done because there are no available treatments for them.

DETECTING CARRIER STATUS

For a number of recessive disorders, it is possible to identify a person who carries the disease-causing gene and may be at risk of transmitting it to a child.

Ethnic groups

Members of some ethnic and racial groups are susceptible to certain recessive disorders. Tay-Sachs disease is found most frequently in Ashkenazi Jews, sickle cell anemia in blacks, thalassemia in people of Mediterranean origin, and cystic fibrosis in whites. If tests show that both members of a couple are carriers of the same abnormal recessive gene, they may want to seek genetic counseling.

Problems of screening

Although screening for genetic disorders is currently done only in specific at-risk groups (blacks for sickle cell anemia and Ashkenazi Jews for Tay-Sachs disease), it may some day be possible to genetically screen the whole population. However, considerable controversy surrounds the idea of establishing widespread screening programs for genetic disorders. Among the major concerns are the age at which a person should be offered testing, and the impact that knowing his or her genetic susceptibilities could have on a person's employment opportunities, health insurance coverage, and choice of a partner. To be effective, screening programs must include both education and counseling.

A common problem
Geneticists believe that everyone carries several defective recessive genes. Unless you have children with a person who has the same abnormal recessive gene, you are unlikely to transmit a disorder to your children.

GENE THERAPY

Medical science may soon make it possible to treat many genetic disorders that are currently untreatable. Cystic fibrosis, for example, may be controlled with regular infusions of healthy genes into an affected person's lungs. Other genetic disorders, including sickle cell anemia, may be cured by replacing defective genes with normal ones.

Diet and genetic disease
When a genetic defect makes the body unable to tolerate certain foods, a restricted diet may relieve many of the symptoms that can develop as a result. For example, people with galactosemia must exclude the milk sugar galactose from their diet.

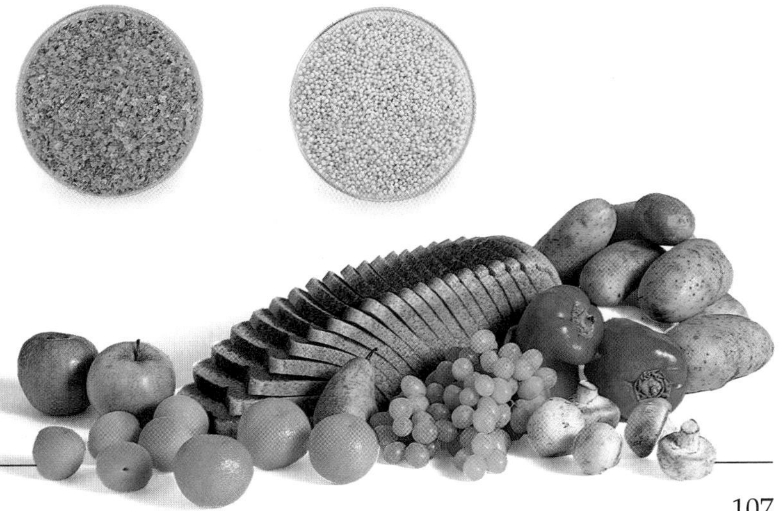

GENETIC ANALYSIS

DOCTORS AND GENETICISTS can use a number of sophisticated laboratory techniques to study genetic diseases and help the many people who have, or are at increased risk of having, an inherited illness. Some tests can diagnose genetic diseases before birth, during childhood, or in adulthood. Other tests can determine if a person is an unaffected carrier of a specific disease-causing gene and at risk of transmitting it to a child.

Analyzing DNA
New methods of analyzing DNA have revolutionized the study and diagnosis of genetic disorders. These methods have allowed scientists to first determine the location of disease-causing genes on human chromosomes and then look for those genes in particular people.

Laboratory study of genetic disorders involves either looking for abnormalities in the body's chemistry caused by faulty genes (biochemical analysis) or looking for defects in the genetic material itself (chromosome or DNA analysis). The biggest advance in the study of genetic diseases in the last 5 years has been in DNA analysis. One of the most powerful of the new diagnostic techniques enables scientists to quickly make millions of copies of specific DNA segments. Such a huge supply of the same piece of DNA makes it possible for them to detect mistakes in individual genes more easily.

BIOCHEMICAL ANALYSIS

Biochemical testing techniques are often used to diagnose genetic disorders that are caused by deficiencies of various enzymes. These disorders are called inborn errors of metabolism (see page 71).

Using biochemical analysis, scientists can measure the levels of different substances in cells as well as the activity of specific enzymes inside cells. Tay-Sachs disease, which is caused by a deficiency or total lack of the enzyme hexosaminidase A, results in a fatal buildup of a poisonous substance in the brain. By measuring the level of activity of the enzyme in a sample of a person's blood, doctors can diagnose Tay-Sachs disease.

A problem inherent in biochemical analysis is the difficulty of obtaining body samples that contain the enzyme or other substance under study. Some proteins produced by defective genes accumulate in the blood or urine. However, many genetic defects show up only in specific cells or tissues. In these cases, a biopsy is performed to remove a small sample of tissue to make a diagnosis possible.

CHROMOSOME ANALYSIS

Chromosome analysis is used primarily to study and detect chromosome abnormalities (see page 60).

Living cells from a person, including those from a fetus, can be examined under a microscope to see if they contain

HOW CHROMOSOMES ARE STUDIED

1 Cells are obtained in a sample of a person's blood; fetal cells are obtained using a prenatal diagnostic technique such as amniocentesis. The cells are then grown in the laboratory.

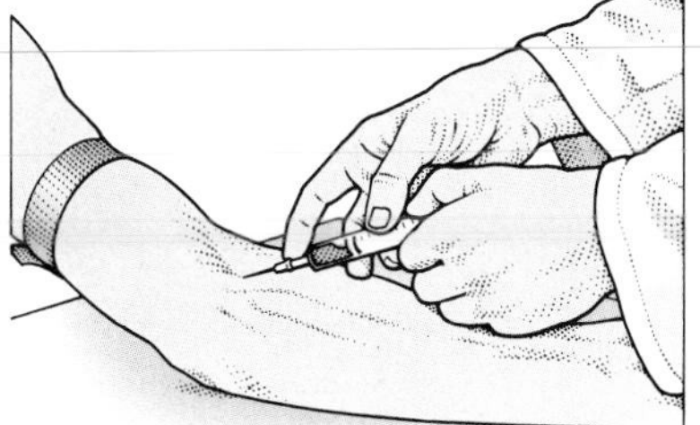

2 When a population of rapidly dividing cells has been obtained, chemicals that stop the cells from dividing further are added at a stage in the division process when the chromosomes are visible.

3 The cells are spread out and stabilized on microscope slides.

4 The chromosomes are stained. Stains are used to produce the characteristic DNA patterns (called banding) of individual chromosomes.

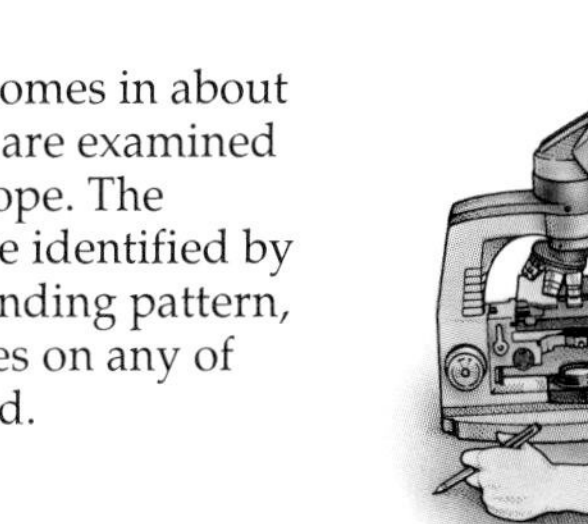

5 The chromosomes in about 10 to 30 cells are examined under a microscope. The chromosomes are identified by their size and banding pattern, and abnormalities on any of them are detected.

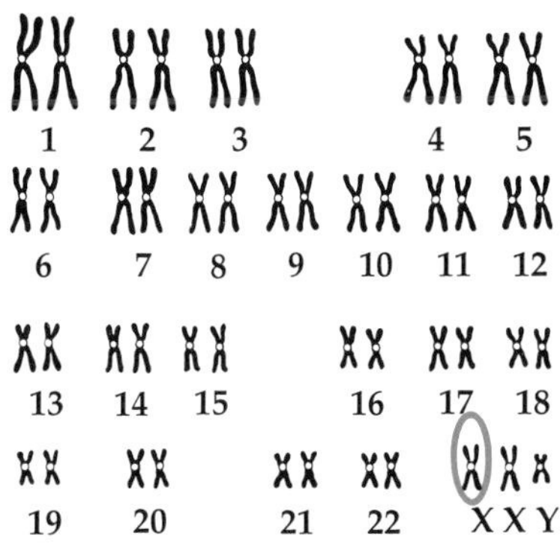

6 To produce an image of a person's chromosomes (a karyotype), a cell is first photographed. The 46 chromosomes are cut from the picture, paired with their matching chromosomes, and the 23 pairs are arranged by number. The karyotype at left is of a person with an extra X sex chromosome.

any chromosome abnormalities. For example, tests can detect the presence of an extra chromosome 21 in a fetus's cells, which is the most common cause of Down's syndrome. In a couple with a history of miscarriages, tests can determine whether either partner has a rearrangement of chromosomes such as a balanced translocation (see page 64). Such a genetic rearrangement can increase the couple's risk of having a child with a chromosome disorder.

Blood is the most common source of cells for study. White blood cells are easy to obtain and they divide quickly (within 48 hours) in the laboratory. For prenatal diagnosis, the most common source of fetal cells is amniotic fluid obtained by amniocentesis (see page 126). It takes an average of 2 to 3 weeks to grow enough fetal cells in the laboratory to study.

DNA ANALYSIS

DNA analysis has been used primarily to identify and study single-gene disorders (see page 70), such as hemophilia and sickle cell anemia. Studying DNA enables doctors to determine if a person is a carrier of a specific X-linked recessive or autosomal (non-sex-linked) recessive disorder. DNA analysis is also used to diagnose many common single-gene disorders in a fetus.

Using genetic probes

Most DNA analysis involves cutting a sample of a person's DNA into millions of pieces, using biological "scissors" called restriction enzymes. The pieces are then examined using other pieces of DNA called genetic probes. Genetic probes can directly detect, for example, a type of gene mutation called a deletion that involves the loss of part of a gene. Approximately 65 percent of boys with Duchenne type muscular dystrophy are missing parts of a particular gene on the X chromosome. Genetic probes can detect this gene defect in a fetus before birth.

WHAT IS A GENETIC PROBE?

A genetic probe is a piece of single-stranded human DNA that has been radioactively labeled. A probe contains a DNA pattern that matches the pattern in the DNA of the person being studied. Some probes can directly detect the presence or absence of specific genes (both normal and abnormal). Other genetic probes are used to identify so-called marker DNA patterns that serve as landmarks for particular disease-causing genes.

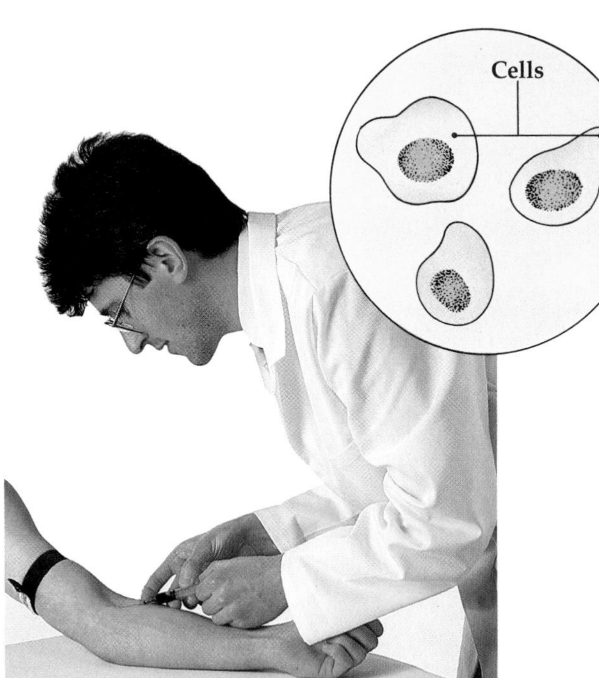

Mass of DNA in test tube

1 DNA is obtained from white blood cells in a sample of a person's blood or from fetal cells obtained by amniocentesis or chorionic villus sampling.

DNA base sequence to be studied

Film being examined

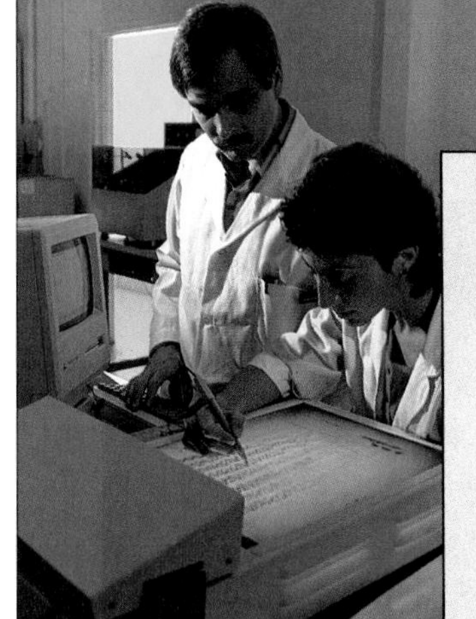

Radiation given off

Gene mutations
The illustration shows the results of DNA tests on four boys using genetic probes to search for parts of the Duchenne type muscular dystrophy gene. Each dark band represents radiation given off when a probe binds to a matching DNA fragment. Column D shows five bands in the DNA of a healthy boy. The other columns represent DNA from boys with the disorder. The absence of some bands or the appearance of bands of a different size indicate mutations in the disease-causing gene.

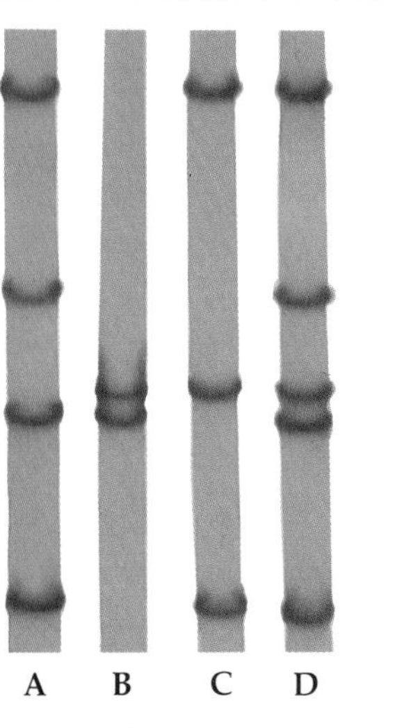

6 If the probe encounters the matching DNA pattern, the probe will bind to that piece of DNA. This binding appears on a film called an autoradiograph as a dark band at a particular location. If the DNA pattern being sought is not present, the probe does not bind, no radiation is given off, and no dark band appears on the film.

2 The DNA is treated with restriction enzymes, the biological "scissors" that can recognize a specific pattern in the DNA and cut it wherever that pattern occurs. The total DNA ends up being cut into millions of pieces of different lengths.

3 These DNA pieces are placed on the surface of a special gel and then separated. The gel is subjected to an electric charge, which causes the pieces to separate according to size.

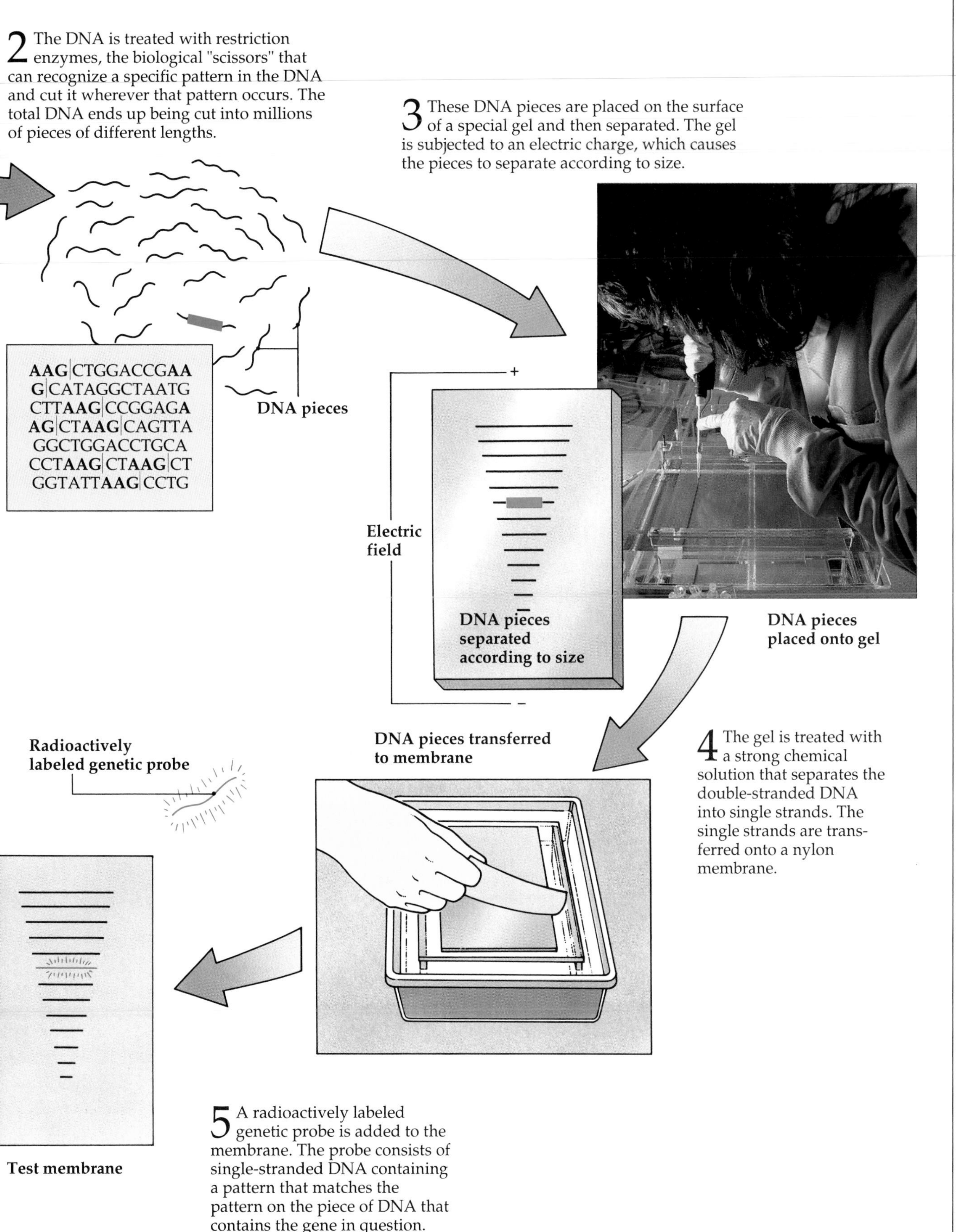

4 The gel is treated with a strong chemical solution that separates the double-stranded DNA into single strands. The single strands are transferred onto a nylon membrane.

5 A radioactively labeled genetic probe is added to the membrane. The probe consists of single-stranded DNA containing a pattern that matches the pattern on the piece of DNA that contains the gene in question.

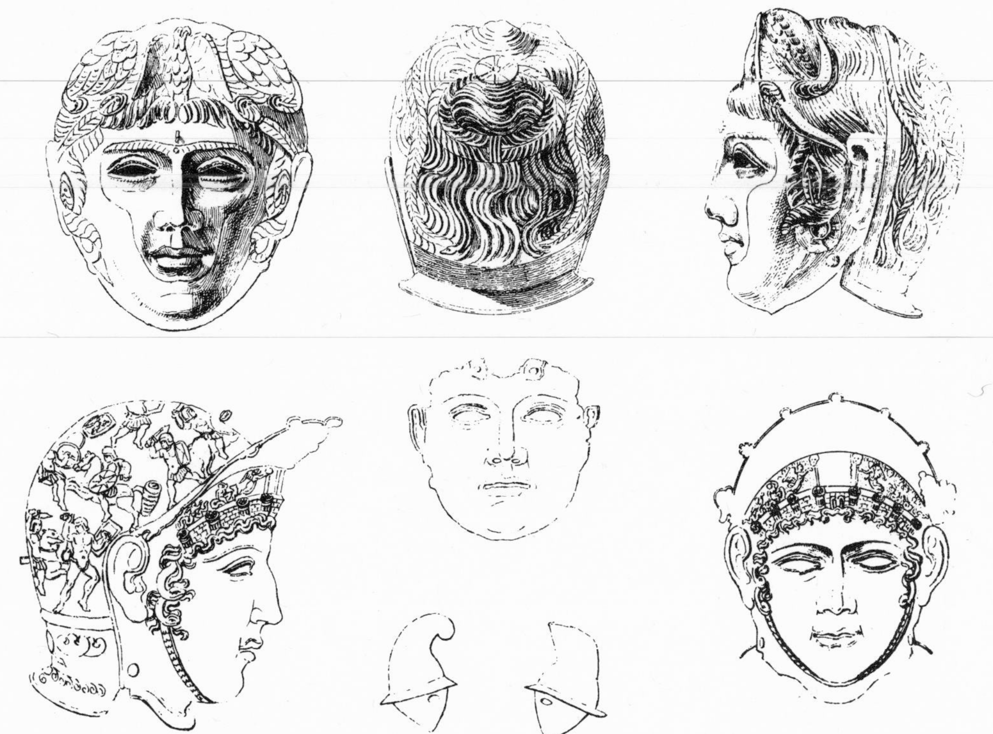

BOOKS FOR FURTHER READING AND SOURCES OF ILLUSTRATIONS

A. Reifenberg, *Ancient Hebrew Arts*. Schocken Books, New York 1950.

C. Blumlein, *Bilder aus dem romisch-germanischen kulturleben*. Verlag Von R. Oldenbourg, Berlin 1918.

M. Wilson, *The Roman toga*. The Johns Hopkins Press, Baltimore.

J. W. Benson, *Time and timetellers*. Robert Hardwicke 1875.

Antiquity and Survival, (the Holy Land) Vol. II no. 2/3. The Hague and Jerusalem 1957.

A. Forestier, *Roman soldier*. A. & C. Black, London 1928.

Edwin R. Goodenough, *Symbolism in the Duva synagogue* (3 vols). Bollinger series xxvii Pantheon Books, New York 1967.

Y. Yadin, *Masada*. Weidenfeld & Nicolson, London 1966.

J. P. V. D. Balsdon, *Life and leisure in ancient Rome*. The Bodley Head, London 1969.

M. Grant, *Gladiators*. Weidenfeld & Nicolson, London 1967.

B. Landström, *The ship*. Allen & Unwin, London 1961.

A. Adams, *Roman antiquities*. Thomas Tegg, London 1836.

L. Casson, *Ships and seamanship in the ancient world*. Princeton Univ. Press 1971.

B. Flower & E. Rosenbaum, *The Roman cookery book*. Harrap, London 1958.

L. Lindenschmit, *Tracht und Bewaffnung des romischen Heeres während der Kaiserzeit*. Brunswick 1882.

H. M. D. Parker, *The Roman legions*. W. Heffer, Cambridge 1961.

G. R. Watson, *The Roman soldier*. Thames & Hudson, London 1969.

W. Ramsay, *Manual of Roman antiquities*. Charles Griffin, London 1898.

W. Smith, *Greek and Roman antiquities*. Taylor and Walton, London 1842.

W. Corswant *A dictionary of life in Bible times* (trans. A. Heathcote). Hodder & Stoughton, London 1960.

Giulio Giannelli (ed.), *The world of ancient Rome*. Putnam, New York 1967.

Webster, *The Roman Imperial army*. A. & C. Black, London 1969.

M. Avi-Yonan, *A history of the Holy Land*. Weidenfeld & Nicolson, London 1969.

V. E. Paoli, *Rome: its people, life and customs*. Longmans, London 1963.

Aegyptisches Museum, Berlin; British Museum, London; Rijksmuseum van Outheiden, Leyden; Israel Maritime Museum, Haifa; Louvre, Paris.

The illustration on this page comes from L. Lindenschmit, Tracht und Bewaffnung des Römischen Heeres während der Kaiserzeit.

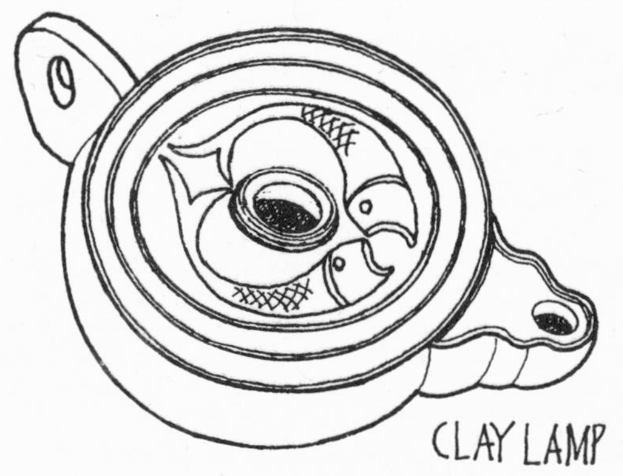
CLAY LAMP

Lew Wallace (1827–1905) wrote *Ben-Hur*, when he was Governor of the new Union State of New Mexico. The novel was first published in 1880. This edition has been edited, annotated and abridged by Robin S. Wright B.A.(Oxon.)

LEW WALLACE

BEN-HUR